Lecture Notes in Computer Science 9596

Commenced Publication in 1973
Founding and Former Series Editors:
Gerhard Goos, Juris Hartmanis, and Jan van Leeuwen

Editorial Board

Colin Johnson · Vic Ciesielski
João Correia · Penousal Machado (Eds.)

Evolutionary and Biologically Inspired Music, Sound, Art and Design

5th International Conference, EvoMUSART 2016
Porto, Portugal, March 30 – April 1, 2016
Proceedings

 Springer

Editors
Colin Johnson
University of Kent
Canterbury
UK

Vic Ciesielski
RMIT University
Melbourne, VIC
Australia

João Correia
University of Coimbra
Coimbra
Portugal

Penousal Machado
University of Coimbra
Coimbra
Portugal

ISSN 0302-9743   ISSN 1611-3349   (electronic)
Lecture Notes in Computer Science
ISBN 978-3-319-31007-7   ISBN 978-3-319-31008-4   (eBook)
DOI 10.1007/978-3-319-31008-4

Library of Congress Control Number: 2016932515

LNCS Sublibrary: SL1 – Theoretical Computer Science and General Issues

Printed on acid-free paper

This Springer imprint is published by Springer Nature
The registered company is Springer International Publishing AG Switzerland

Preface

EvoMUSART 2016—the 5th International Conference and the 13th European event on Biologically Inspired Music, Sound, Art and Design—took place March 30 to April 1, 2016 in Porto, Portugal. It brought together researchers who use biologically inspired computer techniques for artistic, aesthetic, and design purposes. Researchers presented their latest work in the intersection of the fields of computer science, evolutionary systems, art, and aesthetics. As always, the atmosphere was fun, friendly, and constructive.

EvoMUSART has grown steadily since its first edition in 2003 in Essex, UK, when it was one of the Applications of Evolutionary Computing workshops. Since 2012 it has been a full conference as part of the evo* co-located events.

EvoMUSART 2016 received 25 submissions. The peer-review process was rigorous and double-blind. The international Program Committee, listed here, was composed of 58 members from 22 countries. EvoMUSART continued to provide useful feedback to authors: Among the papers sent for full review, there were on average 3.12 reviews per paper. It also continued to ensure quality by keeping acceptance rates low: ten papers were accepted for oral presentation (40 % acceptance rate), and six for poster presentation (24 % acceptance rate).

This volume of proceedings collects the accepted papers. As always, the EvoMUSART proceedings cover a wide range of topics and application areas, including: generative approaches to music, graphics, game content, and narrative; music information retrieval; computational aesthetics; the mechanics of interactive evolutionary computation; and the art theory of evolutionary computation.

We thank all authors for submitting their work, including those whose work was not accepted for presentation. As always, the standard of submissions was high, and good papers had to be rejected.

The work of reviewing is done voluntarily and generally without official recognition from the institutions where reviewers are employed. Nevertheless, good reviewing is essential to a healthy conference. Therefore we particularly thank the members of the Program Committee for their hard work and professionalism in providing constructive and fair reviews.

EvoMUSART 2016 was part of the evo* 2016 event, which included three additional conferences: evoGP 2016, evoCOP 2016, and evoApplications 2016. Many people helped to make this event a success.

We thank the Câmara Municipal do Porto (the city hall) and Turismo do Porto involvement in the event. We thank the local organizing team of Penousal Machado and Ernesto Costa, from the University of Coimbra.

We thank Marc Schoenauer (Inria Saclay, Île-de-France), for continued assistance in providing the MyReview conference management system. We thank Pablo García Sánchez (University of Granada) for evo* publicity and website services.

We want to especially acknowledge our invited speakers: Richard Forsyth and Kenneth Sörensen.

Last but certainly not least, we especially want to express a heartfelt thanks to Jennifer Willies and the Institute for Informatics and Digital Innovation at Edinburgh Napier University. Ever since its inaugural meeting in 1998 this event has relied on her dedicated work and continued involvement and we do not exaggerate when we state that without her, evo* could not have achieved its current status.

April 2016

Colin Johnson
Vic Ciesielski
João Correia
Penousal Machado

Organization

EvoMUSART 2016 was part of evo* 2016, Europe's premier co-located events in the field of evolutionary computing, which also included the conferences euroGP 2016, evoCOP 2016, evoBIO 2016, and evoApplications 2016.

Organizing Committee

Conference Chairs

Colin Johnson   University of Kent, UK
Vic Ciesielski   RMIT, Australia

Publication Chair

João Correia   University of Coimbra, Portugal

Local Chair

Penousal Machado   University of Coimbra, Portugal

Program Committee

Pedro Abreu   University of Coimbra, Portugal
Dan Ashlock   University of Guelph, Canada
Peter Bentley   University College London, UK
Eleonora Bilotta   University of Calabria, Italy
Tim Blackwell   Goldsmiths College, University of London, UK
Andrew Brown   Griffith University, Australia
Adrian Carballal   University of A Coruna, Spain
Amílcar Cardoso   University of Coimbra, Portugal
Vic Ciesielski   RMIT, Australia
João Correia   University of Coimbra, Portugal
Pedro Cruz   University of Coimbra, Portugal
Palle Dahlstedt   Göteborg University, Sweden
Eelco den Heijer   Vrije Universiteit Amsterdam, The Netherlands
Alan Dorin   Monash University, Australia
Arne Eigenfeldt   Simon Fraser University, Canada
Jonathan Eisenmann   Ohio State University, USA
José Fornari   NICS/Unicamp, Brazil
Marcelo Freitas Caetano   INESC TEC, Portugal
Philip Galanter   Texas A&M College of Architecture, USA
Pablo Gervás   Universidad Complutense de Madrid, Spain
Andrew Gildfind   Google Inc., Australia
Gary Greenfield   University of Richmond, USA
Carlos Grilo   Instituto Politécnico de Leiria, Portugal

Contents

Computer-Aided Musical Orchestration Using an Artificial Immune System

José Abreu[1]([⊠]), Marcelo Caetano[2], and Rui Penha[1,2]

[1] Faculty of Engineering, University of Porto, Porto, Portugal
{ee10146,ruipenha}@fe.up.pt
[2] Sound and Music Computing Group, INESC TEC, Porto, Portugal
mcaetano@inesctec.pt

Abstract. The aim of computer-aided musical orchestration is to find a combination of musical instrument sounds that approximates a target sound. The difficulty arises from the complexity of timbre perception and the combinatorial explosion of all possible instrument mixtures. The estimation of perceptual similarities between sounds requires a model capable of capturing the multidimensional perception of timbre, among other perceptual qualities of sounds. In this work, we use an artificial immune system (AIS) called opt-aiNet to search for combinations of musical instrument sounds that minimize the distance to a target sound encoded in a fitness function. Opt-aiNet is capable of finding multiple solutions in parallel while preserving diversity, proposing alternative orchestrations for the same target sound that are different among themselves. We performed a listening test to evaluate the subjective similarity and diversity of the orchestrations.

1 Introduction

Orchestration refers to composing music for an orchestra [12]. Initially, orchestration was simply the assignment of instruments to pre-composed parts of the score, which was dictated largely by availability of resources, such as what instruments there are and how many of them [10,12]. Later on, composers started regarding orchestration as an integral part of the compositional process whereby the musical ideas themselves are expressed [18]. Compositional experimentation in orchestration arises from the increasing tendency to specify instrument combinations to achieve desired effects, resulting in the contemporary use of timbral combinations [15,18]. The development of computational tools that aid the composer in exploring the virtually infinite possibilities resulting from the combinations of musical instruments gave rise to computer-aided musical orchestration (CAMO) [3–6,11,17,18]. Most of these tools rely on searching for combinations of musical instrument sounds from pre-recorded datasets to approximate a given target sound. Early works [11,17,18] resorted to spectral analysis followed by subtractive spectral matching.

Psenicka [17] describes SPORCH (SPectral ORCHestration) as "a program designed to analyze a recorded sound and output a list of instruments, pitches, and dynamic levels that when played together create a sonority whose timbre

© Springer International Publishing Switzerland 2016
C. Johnson et al. (Eds.): EvoMUSART 2016, LNCS 9596, pp. 1–16, 2016.
DOI: 10.1007/978-3-319-31008-4_1

and quality approximate that of the analyzed sound." The method keeps a database of spectral peaks estimated from either the steady state or the attack (for nonpercussive and percussive sounds, respectively) of musical instrument sounds organized according to pitch, dynamic level, and playing technique such as *staccato* and *vibrato*. The algorithm iteratively subtracts the spectral peaks of the best match from the target spectrum aiming to minimize the residual spectral energy in the least squares sense. The iterative procedure requires little computational power, but the greedy algorithm restricts the exploration of the solution space, often resulting in suboptimal solutions because it only fits the best match per iteration. Hummel [11] approximates the spectral envelope of phonemes as a combination of the spectral envelopes of musical instrument sounds. The method also uses a greedy iterative spectral subtraction procedure. The spectral peaks are not considered when computing the similarity between target and candidate sounds, disregarding pitch among other perceptual qualities. Rose and Hetrik [18] use singular value decomposition (SVD) to perform spectral decomposition and spectral matching using a database of averaged DFTs of musical instrument sounds containing different pitches, dynamic levels, and playing techniques. SVD decomposes the target spectrum as a weighted sum of the instruments present in the database, where the weights reflect the match. Besides the drawbacks from the previous approaches, SVD can be computationally intensive even for relatively small databases. Additionally, SVD sometimes returns combinations that are unplayable such as multiple simultaneous notes on the same violin, requiring an additional procedure to specify constraints on the database that reflect the physical constraints of musical instruments and of the orchestra.

The concept of timbre lies at the core of musical orchestration. Yet, timbre perception is still only partially understood [1,9,13,15,16]. The term timbre encompasses auditory attributes, perceptual and musical issues, covering perceptual parameters not accounted for by pitch, loudness, spatial position, duration, among others [13,15]. Nowadays, timbre is regarded as both a multidimensional set of sensory attributes that quantitatively characterize the ways in which sounds are perceived to differ and the primary vehicle for sound source recognition and identification [15]. McAdams and Bruno [15] wrote that "instrumental combinations can give rise to new timbres if the sounds are perceived as blended, and timbre can play a role in creating and releasing musical tension." Consequently, the goal of CAMO is to find an instrument combination that best approximates the target timbre rather than the target spectrum [19].

To overcome the drawbacks of subtractive spectral matching, Carpentier and collaborators [3–6,19] search for a combination of musical instrument sounds whose timbral features best match those of the target sound. This approach requires a model of timbre perception to describe the timbre of isolated sounds, a method to estimate the timbral result of an instrument combination, and a measure of timbre similarity to compare the combinations and the target. Multidimensional scaling (MDS) of perceptual dissimilarity ratings [1,9,13,16] provides a set of auditory correlates of timbre perception that are widely used to model timbre perception of isolated musical instrument sounds. MDS spaces are obtained by equating distance measures to timbral (dis)similarity ratings.

In metric MDS spaces, the distance measure directly allows timbral comparison. Models of timbral combination [6,12] estimate these features for combinations of musical instrument sounds.

Carpentier and collaborators [3–6,19] consider the search for combinations of musical instrument sounds as a constrained combinatorial optimization problem [5]. They formulate CAMO as a variation of the knapsack problem where the aim is to find a combination of musical instruments that maximizes the timbral similarity with the target constrained by the capacity of the orchestra (i.e., the database). The binary allocation knapsack problem can be shown to be NP-complete so it cannot be solved in polynomial time. They explore the vast space of possible instrument combinations with a genetic algorithm (GA) that optimizes a fitness function which encodes timbral similarity between the candidate instrument combinations and the target sound. GAs are metaheuristics inspired by the Darwinian principle of *survival of the fittest*. The GA maintains a list of individuals that represent the possible combinations of instruments. These individuals evolve towards optimal solutions by means of crossover, mutation, and selection. Crossover and mutation are responsible for introducing variations in the current population and promoting the exploration and exploitation of the search space. Selection guarantees that the fittest individuals are passed to the next generation gradually converging to optimal regions of the search space. The major drawback of this approach arises from the loss of diversity inherent in the evolutionary search performed with GAs. In practice, the loss of diversity results in only one solution that commonly corresponds to a local optimum because GAs cannot guarantee to return the global optimum (i.e., the best solution). Moreover, running the GA multiple times with the same parameters commonly results in different solutions. Carpentier *et al.* [5] use a combination of local search and constraint strategies to circumvent the issues resulting from loss of diversity.

In this work, we use an artificial immune system (AIS) called opt-aiNet [7] to search for multiple combinations of musical instrument sounds whose timbral features match those of the target sound. Inspired by immunological principles, opt-aiNet returns multiple good quality solutions in parallel while preserving diversity. The intrinsic property of maintenance of diversity allows opt-aiNet to return all the optima (global and local) of the fitness function being optimized upon convergence, which translates as orchestrations that are all similar to the target yet different from one another. The AIS provides the composer with multiple choices when orchestrating a sound instead of searching for one solution constrained by choices defined *a priori*. Therefore, our work can expand the creative possibilities of CAMO beyond what the composer initially imagined.

The remainder of this paper is organized as follows. The next section presents an overview of our approach to CAMO. Then we describe the immunological approach to CAMO. Next we present the experiment we performed followed by the evaluation. The evaluation comprises similarity and diversity using objective measures and the subjective ratings from a listening test. We present and discuss the results, followed by conclusions and perspectives.

2   Computer-Aided Musical Orchestration (CAMO)

2.1   Overview

Figure 1 shows an overview of our approach. The sound database is used to build a feature database, which consists of acoustic features calculated for all sounds prior to the search for orchestrations. The same features are calculated for the target sound being orchestrated. The fitness function uses these features to estimate the similarity between combinations of features from sounds in the database and those of the target sound. The AIS is used to search for combinations that approximate the target sound, called orchestrations. Each orchestration is a list of sounds from the sound database, which contains sounds with various lengths. A phase vocoder is used to time-stretch or compress each sound from an orchestration to the average duration to ensure they all start and end at the same time when played together. The graphic interface (GUI) displays information about the solution set and allows the user to play the target sound and the orchestrations.

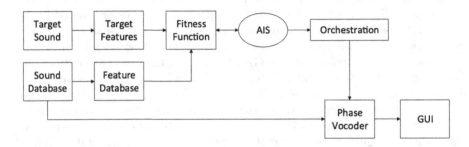

Fig. 1. Overview of the modules that compose the developed orchestration system

2.2   Sound Database

The sound database used in this work contains musical instrument sounds from the RWC Music Database [8] available to compose the orchestrations. In total, there are 1439 sounds from 13 instruments played with 3 dynamics, *forte, mezzo forte* and *piano*. The instruments are *violin, viola, cello, contrabass, trumpet, trombone, tuba, french horn, english horn, oboe, bassoon, clarinet,* and *flute*. For each file the values of the sound features described in the next section were computed and stored.

2.3   Feature Database

Traditionally, timbre is considered as the set of attributes whereby a listener can judge that two sounds are dissimilar using any criteria other than pitch, loudness, or duration [15]. Therefore, we consider pitch, loudness, and duration separately from timbre dimensions. The features used are fundamental frequency f_0 (pitch), frequency and amplitude of the contribution spectral peaks P, loudness λ, spectral centroid μ, and spectral spread σ. The spectral centroid μ captures brightness while the spectral spread σ correlates with the third dimension of MDS timbre spaces [1,9,13,16]. The RMS energy was also calculated for each sound and the duration is equalized later with a phase vocoder. All musical instrument sounds used are sustained (i.e., nonpercussive) with attack times longer than 250 ms and duration of 1 s or more. The calculation of the features is performed for short-term frames between 250 ms and 750 ms and then averaged because the signal is considered stable in that region.

Fundamental Frequency. The f_0 of all sounds $s(i)$ in the database is estimated with Swipe [2].

Contribution Spectral Peaks. The spectral peaks considered are those whose spectral energy (amplitude squared) is at most 35 dB below the maximum. They are estimated with the MIR Toolbox [14] and stored as a vector with the pairs $\{a(n), f(n)\} \in s(i)$. The spectral peaks $a(n)$ are used to compute the *contribution* spectral peaks $P(i,n)$, which are the spectral peaks from the *selected* sound $s(i)$ that are common to the spectral peaks of the *target* sound s^T. Equation (1) shows the calculation of $P(i,n)$ as

$$P(i,n) = \begin{cases} a^i(n) & \text{if } (1+\delta)^{-1} \leq f^i(n)/f^T(n) \leq 1+\delta \\ 0 & \text{otherwise} \end{cases} \quad (1)$$

where $a^i(n)$ is the amplitude and $f^i(n)$ is the frequency of the main spectral peak of the *selected* sound, and $f^T(n)$ is the frequency of the *target* sound. Figure 2 illustrates the computation of spectral peak similarity between the *target* sound and a *selected* sound. Spectral peaks are represented as spikes with amplitude $a(n)$ at frequency $f(n)$ where n is the index of the peak. The frequencies $f^T(n)$ of the peaks of the *target* sound are used as reference. Whenever the *selected* sound contains a peak in a region δ around $f^T(n)$, the amplitude $a^i(n)$ of the peak at frequency $f^i(n)$ of the *selected* sound is kept at position n of the contribution vector $P(n)$. In this work $\delta = 0.025$.

Loudness. Loudness $\lambda(i)$ is calculated as

$$\lambda(i) = 20 \log_{10} \left(\sum_n a(n) \right), \quad (2)$$

where $a(n)$ are the amplitudes at frequencies $f(n)$ and i is the sound index.

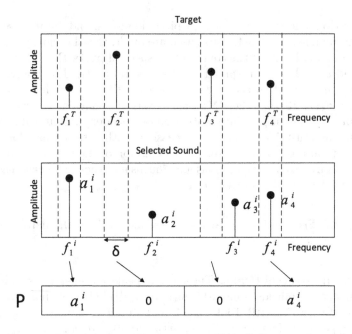

Fig. 2. Construction of the contribution vector $P(i, n)$. See text for explanation.

Spectral Centroid. The spectral centroid $\mu(i)$ is calculated as

$$\mu(i) = \sum_n f(n) \frac{|a(n)|^2}{\sum_n |a(n)|^2}. \tag{3}$$

Spectral Spread. The spectral spread $\sigma(i)$ is calculated as

$$\sigma(i) = \sum_n (f(n) - \mu)^2 \frac{|a(n)|^2}{\sum_n |a(n)|^2}. \tag{4}$$

2.4 Pre-processing

Prior to the search for orchestrations of a given target sound s^T, the entire sound database S is reduced to a subset S^T of sounds that will be effectively used to compose orchestrations for s^T. All the sounds whose contribution vector $P(i, n)$ is all zeros are eliminated because these do not have any contribution spectral peaks. Similarly, all the sounds whose f_0 is lower than f_0^T are eliminated because any partials lower than f_0^T have a negative impact on the final result. Partials that are higher than all $P^T(n)$ have a negligible effect and are not considered.

2.5 Representation

An orchestration is a list of sounds $S(i)$ that, when played together, should approximate the target sound s^T. Thus orchestrations are represented as $S(i) =$

$\{s(1), s(2), \ldots, s(i), \ldots, s(I)\}, \quad \forall s(i) \in S^T$. In practice, $S(i)$ has I sounds, and each sound $s(i)$ corresponds to a note of a given instrument played with a dynamic level. Zero indicates no instrument.

2.6 Combination Functions

The sounds $s(i)$ in an orchestration $S(i)$ should approximate the target s^T when played together. Therefore, the combination functions estimate the values of the spectral features of $S(i)$ from the features of the isolated sounds $s(i)$ normalized by the RMS energy $e(i)$ [6]. The combination functions for the spectral centroid $\mu(i)$, spectral spread $\sigma(i)$, and loudness $\lambda(i)$ are given respectively by

$$\mu(S(i)) = \frac{\sum_i^I e(i)\mu(i)}{\sum_i^I e(i)} \tag{5}$$

$$\sigma(S(i)) = \sqrt{\frac{\sum_i^I e(i)\left(\sigma^2(i) + \mu^2(i)\right)}{\sum_i^I e(i)} - \mu^2(S(i))} \tag{6}$$

$$\lambda(S(i)) = 20\log_{10}\left(\sum_i^I \frac{1}{N}\sum_n a(i,n)\right) \tag{7}$$

The estimation of the contribution spectral peaks of the combination $P(S(i), n)$ uses the contribution vectors $P(i, n)$ of the sounds $s(i)$ in $S(i)$ as

$$P(S(i), n) = \left\{\max_{i \in I}[P(i,1)], \max_{i \in I}[P(i,2)], \cdots, \max_{i \in I}[P(i,N)]\right\} \tag{8}$$

2.7 Fitness Function

The fitness value $F(S(i))$ of an orchestration $S(i)$ is calculated as

$$F(S(i)) = -\sum_y \alpha(y)D_y \tag{9}$$

where $\alpha(y)$ are the weights that establish the relative importance of the absolute distances D_y. Each D_y, in turn, measures the difference between the features from the target sound s^T and the candidate orchestration $S(i)$ as follows

$$D_\mu = \frac{|\mu(S(i)) - \mu(s^T)|}{\mu(s^T)} \tag{10}$$

$$D_\sigma = \frac{|\sigma(S(i)) - \sigma(s^T)|}{\sigma(s^T)} \tag{11}$$

$$D_\lambda = \frac{|\lambda\left(S\left(i\right)\right) - \lambda\left(s^T\right)|}{\lambda\left(s^T\right)} \tag{12}$$

The use of Eqs. (10) and (11) is specially suited for CAMO problems as these measures are more sensitive for lower frequencies, a fact that is desired considering human perception of sound. The distance between the contribution vector of the target sound $P\left(s^T, n\right)$ and the contribution vector of the orchestration $P\left(S\left(i\right), n\right)$ is calculated as

$$D_P = 1 - \cos(P\left(S\left(i\right), n\right), P\left(s^T, n\right)). \tag{13}$$

The weights used in this work are

$$\alpha = \{0.1, 0.1, 0.2, 0.6\}. \tag{14}$$

The aim of CAMO is to find $S\left(i\right)$ that is close to s^T. Thus we want to minimize the distances D_y that comprise F. The negative sign in Eq. (9) makes the fitness value of all combinations negative so maximizing F approaches zero and minimizes the distance from s^T. However, the fitness landscape of $F\left(S\left(i\right)\right)$ depends on the combinations $S\left(i\right)$ in S^T, giving rise to a complex space which requires an optimization method to find solutions.

3   Immune Inspired Musical Orchestration

The work of Carpentier *et al.* [3–6,19] represents a paradigm shift in CAMO. Prior to their work, most approaches [11,17,18] used greedy search procedures based on subtractive spectral matching. Their contribution is twofold, the use of perceptually related features to measure timbral similarity and the use of constrained combinatorial optimization to search for combinations of musical instruments whose features approach those of the target sound. Timbral similarity is encoded in a fitness function such that better combinations present higher fitness values. The aim is to find combinations that correspond to maxima of the fitness function as illustrated in Fig. 3. The surfaces represent the fitness function, which might have multiple peaks, and the black dots represent the fitness values of specific instrument combinations. The combinatorial explosion resulting from the exhaustive search of all possible combinations requires heuristics to find a solution in less time.

Carpentier *et al.* [3–6,19] use GAs to perform the search due to their ability to perform exploration and exploitation of the search space. Exploration is responsible for looking for new promising regions of the search space (peaks of the fitness function) and exploitation climbs the peaks looking to improve the current candidate solutions. However, the standard GA suffers from loss of diversity upon convergence, which results in only one solution corresponding to one peak is returned by the GA as shown in Fig. 3a. The stochastic nature of the search procedure does not guarantee that the global optimum is found, often getting stuck in local optima. Additionally, running the GA multiple times with

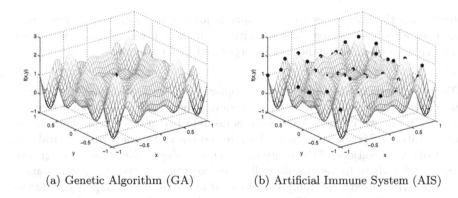

(a) Genetic Algorithm (GA)   (b) Artificial Immune System (AIS)

Fig. 3. Multimodal function optimization. The figure illustrates a fitness function with multiple optima. Part (a) shows that the GA finds only one optimum. Part (b) shows the ability of the AIS to find all the optima of the fitness function.

the same parameters commonly results in different solutions corresponding to different peaks of the fitness function.

We propose an immune inspired approach to CAMO instead. We use an artificial immune system (AIS) called opt-aiNet [7] to perform the search. Figure 3b illustrates the ability of opt-aiNet to find all optima of the fitness function preserving diversity. The ability to maintain diversity translates as solutions that are different from one another.

3.1   Opt-aiNet: An Artificial Immune System for Optimization

Inspired by the natural immune system, De Castro and Timmis [7] developed an artificial immune system (AIS) called opt-aiNet for multimodal optimization problems, which typically present several possible solutions as optima of the fitness function. Opt-aiNet uses the immunological principles of clonal expansion, mutation and suppression to evolve a population of antibodies in an immune network. Opt-aiNet combines local and global search to locate and maintain multiple optima of the fitness function in parallel while preserving diversity of the solutions. This means that opt-aiNet can find a set of good candidates for the solution of the optimization problem that are different from one another.

Each network cell (antibody) is represented as a vector whose fitness is measured with a fitness function. Additionally, the similarity among antibodies is called affinity, and a high affinity means that the antibodies are similar. Affinity is measured with a distance metric such as the Euclidean distance. The antibodies are initialized at random to explore the search space. Some high fitness antibodies are selected and cloned based on their fitness value, the higher the fitness, the higher the number of clones and vice-versa. The clones generated suffer a mutation inversely proportional to their fitness and a number of high fitness clones is maintained in the network as memory. Then, the affinity among the remaining antibodies is determined. Maintenance of diversity is achieved by eliminating the antibodies whose affinity is lower than a given threshold from the network while keeping the ones with the highest fitness. Finally, a number of newly generated antibodies are incorporated into the network.

Discrete Search Space. Originally, opt-aiNet [7] was designed to optimize functions of continuous variables, performing the search in continuous vector spaces. In our work, the search space is discrete because the representation of orchestrations $S(i)$ is a vector of discrete indices i of sounds in the database. Most of the operations of the continuous version of opt-aiNet work as originally intended for discrete vectors as well. The exception is the original mutation operator which used a continuous random variable to add a small perturbation to the vectors being mutated. Thus we adapted the mutation operator for discrete vectors using a probability of mutation to determine if the vector will undergo mutation. The probability of mutation p_m is calculated as

$$p_m = \exp(-\gamma\hat{F}) \tag{15}$$

where γ is a constant and $\hat{F}$ is the normalized fitness value of the combination vector $S(i)$ being mutated. For each index i, a uniform random variable $u(0,1)$ will determine if the corresponding sound $s(i)$ is replaced by another sound from S^T. If $u(0,1) < p_m$ then a new $i \in S^T$ is chosen from another uniform distribution. Here we set $\gamma = 1.2$. The suppression operation discards cells that have affinity values below a given threshold. In this work, the affinity between two antibodies is the distance between vectors whose components are calculated using Eqs. (10)–(13).

3.2   Phase Vocoder

The orchestrations found by the AIS are created as combinations of the sounds from the database, which have different temporal duration. The focus on timbral similarity requires to equalize the other dimensions of sound perception, including the duration of the sounds. The Phase Vocoder (PV)[1] can manipulate the pitch and duration independently, allowing to create combinations of sounds from the database with the same duration while preserving the pitch and other perceptual features.

3.3   Graphical Interface

Figure 4 shows the graphical interface (GUI) that displays the orchestrations proposed by the AIS. The GUI allows to play the orchestrations and shows the instruments comprised in them. The spectrum of the target sound and the orchestrations can also be viewed.

---

[1] http://labrosa.ee.columbia.edu/matlab/pvoc/.

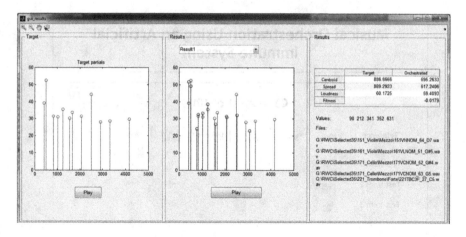

Fig. 4. Graphical interface (GUI).

4   Evaluation

The aim of the evaluation was to investigate the similarity and diversity of the orchestrations proposed by our system. The quality of a solution depends on how similar it is to the target sound. We want all the solutions proposed by the system to be as close to the target sound as possible. However, diversity is also important. Multiple solutions should be different from one another to represent alternatives, giving the user options to choose from. Therefore, we evaluate the similarity and the diversity of the orchestrations proposed by the system and compare them with an implementation of a genetic algorithm.

We performed subjective and objective evaluations for similarity and diversity. The subjective evaluation consists of a listening test, and the objective evaluation uses distance measures. For the listening test, we selected 10 target sounds, listed in Table 1. These sounds were chosen according to two criteria, *temporal variation* and *harmonicity*. Target sounds that have high temporal variation will tend to be more challenging to orchestrate because we use the average value of the features. Similarly, target sounds that are less harmonic will be more challenging to orchestrate with musical instrument sounds because all the musical instruments in the database used to find orchestrations are very close to

Table 1. Target sounds used in the listening test. The table contains an informal estimation of the degree of temporal variation (TempVar) and the degree of harmonicity (Harm) of each target sound as low (L) or high (H).

Target	Horn	Synth	Tbreed	Ahh	Harp	Didger	Eleph	Frog	Scream	Gong
TempVar	L	L	L	L	L	L	H	H	H	H
Harm	L	H	L	H	H	H	L	L	L	L

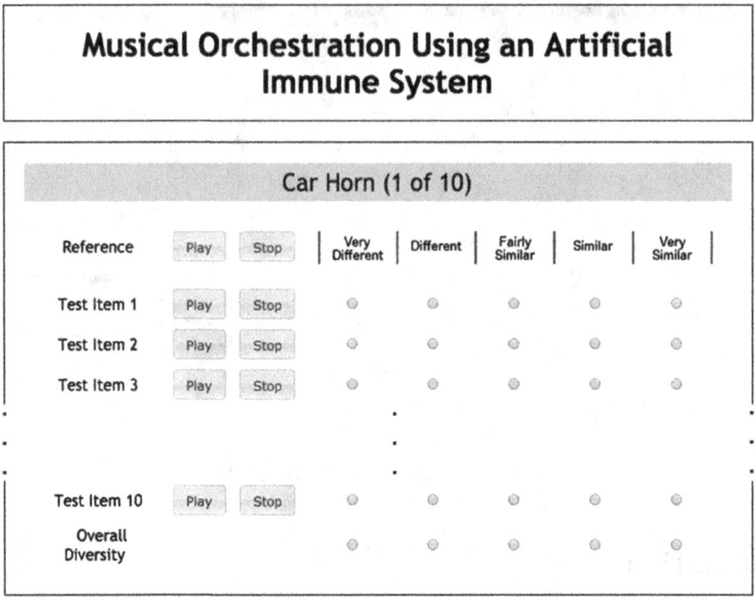

Fig. 5. Illustration of the listening test

harmonic. The listening test shown in Fig. 5 was run online[2]. The target sound is marked as *reference*, and the orchestrations are *test item*. After playing the sounds multiple times if necessary, the listeners were asked to assess the *subjective similarity* between the reference and each test item using the following scale: *very different, different, fairly similar, similar,* and *very similar*. Finally, the listeners were asked to judge the *overall diversity* of the test items using the same scale. In total, 23 people took the test.

The objective evaluation of similarity uses the fitness values of the solutions, while the objective evaluation of diversity uses Eq. (16), which quantifies the number of sounds in common between the orchestrations present in the set as:

$$div = 1 - \frac{k}{I}, \tag{16}$$

where k is the number of common sounds and I the total number of sounds in an orchestration. This equation quantifies the objective diversity as a value between 0 and 1, where 0 corresponds to minimum diversity and 1 corresponds to maximum diversity.

5 Results and Discussion

5.1 Subjective Evaluation

The results obtained for the subjective evaluation are shown in Table 2. Subjective similarity varies from 1 to 5 with higher values corresponding to

---

[2] Access http://goo.gl/weHaHI to see the test and listen to the sounds.

Table 2. Results of the subjective evaluation.

Target	Similarity	Diversity	Target	Similarity	Diversity
Horn	2.6±0.6	3.4±0.9	Didger	2.1±0.3	3.7±1.1
Synth	1.8±0.3	3.9±1.1	Eleph	2.3±0.2	3.3±1.1
Tbreed	2.0±0.3	3.3±1.3	Frog	1.2±0.1	3.4±1.7
Ahh	2.2±0.5	3.6±1.1	Scream	2.7±0.2	2.7±1.2
Harp	2.3±0.2	3.3±1.0	Gong	2.3±0.3	3.4±1.0

orchestrations that are more similar to the target. The value 1 corresponds to the option *very different* in the listening test and the value 5 corresponds to the option *very similar*. Subjective diversity also varies from 1 to 5 with higher values indicating a more diverse set of orchestrations. The value 1 corresponds to the option *very similar* in the listening test and the value 5 corresponds to the option *very different*.

Subjective Similarity. The target sound *Scream* received the highest score for subjective similarity, followed by *Horn*. Table 1 indicates that *Scream* presents high temporal variation and low harmonicity, while *Horn* presents both low temporal variation and harmonicity. We expected sounds with high harmonicity and low temporal variation such as *Synth*, *Ahh*, and *Didger* to render orchestrations that would be considered more similar than the others because the sounds in the database are stable notes from harmonic musical instruments.

The target sound *Frog* received the lowest score for subjective similarity. The croaking of a frog is characterized mostly by the temporal modulations than spectral features, thus we expected *Frog* to be a particularly challenging target sound to orchestrate with sustained notes from musical instruments.

Subjective Diversity. Most subjective diversity scores were above 3 on average (with the exception of *Scream*), corresponding to assessments between *Fairly Similar* and *Different*. This result is an indication that the AIS is capable of returning multiple orchestrations in parallel that correspond to alternative combinations of sounds.

5.2   Objective Evaluation

The evaluation of objective similarity uses the fitness values of the orchestrations while the evaluation of objective diversity uses Eq. (16). Table 3 shows the results for the objective evaluation. Fitness is always negative and the closer to zero the smaller the distance from the target. The objective diversity values vary from 0 to 1 and the higher they are the more diverse the set of orchestrations because they have fewer sounds in common.

Table 3. Objective evaluation. The table shows the fitness (Fit) and Diversity (Div) for the artificial immune system (AIS) and genetic algorithm (GA).

Target	Fit AIS (x10^{-3})	Fit GA (x10^{-3})	Div AIS	Div GA
Horn	−24.3±3.5	−31.7±3.6	0.78±0.16	0.82±0.20
Synth	−14.8±3.4	−22.5±9.3	0.84±0.13	0.83±0.15
Tbreed	−53.0±5.0	−66.8±5.8	0.82±0.15	0.79±0.16
Ahh	−45.6±3.6	−54.0±3.9	0.84±0.14	0.78±0.17
Harp	−23.9±4.9	−35.7±7.6	0.80±0.15	0.78±0.18
Didger	−16.5±3.1	−18.5±1.8	0.64±0.16	0.62±0.18
Eleph	−64.5±4.1	−74.3±7.4	0.86±0.13	0.80±0.17
Frog	−154.3±7.8	−166.8±13.5	0.78±0.15	0.76±0.19
Scream	−58.6±7.4	−55.0±4.2	0.85±0.14	0.64±0.16
Gong	−50.3±5.8	−65.9±12.0	0.82±0.15	0.72±0.14

Objective Similarity. We compare the results obtained by the AIS opt-aiNet with the results obtained using a standard GA. We ran the GA 10 times and stored the fitness value of the best classified solution after each run. The results shown in Table 3 are the average fitness values obtained in the 10 executions of the GA and the average fitness values obtained in a single run of the AIS. The AIS returns multiple solutions, so we averaged the values of the top 10 orchestrations ranked by fitness value. In general, Table 3 shows that the fitness for the AIS is closer to zero than the fitness for the GA, indicating better matches. However, the results of the objective similarity evaluation in Table 3 do not reflect the subjective similarity from Table 2. For example, the orchestrations that have the best fitness values are *Synth* and *Didger* but these target sounds were not considered to render the closest orchestrations. Therefore, the perceptual significance of the fitness function remains to be investigated.

Objective Diversity. The objective diversity was computed using Eq. (16). The diversity for the AIS was calculated from the top 10 ranked solutions from a single run, while the diversity for the GA was calculated after 10 runs. Table 3 indicates that a single run of the AIS returns a set of orchestrations with objective diversity comparable with multiple runs of the GA.

6   Conclusions and Perspectives

We proposed to use an artificial immune system (AIS) to find combinations of sounds from a database that approach a target sound. The AIS is capable of finding multiple combinations that are good candidate solutions while preserving diversity, contrary to the standard GA. We evaluated the objective and subjective similarity of the orchestrations returned by the AIS to 10 target sounds and compared the results with a standard GA. Similarly, we evaluated the objective and subjective diversity of 10 solutions found by the AIS and compared with 10 independent runs of the GA. The orchestrations found by the AIS presented similarity and diversity comparable to running the GA multiple times to obtain different orchestrations.

We focused on spectral features of sounds, neglecting the inherent temporal aspect of sound perception. The attack time is the most salient feature in dissimilarity studies and should be considered when searching for orchestrations. Orchestrating percussive sounds such as the *Gong* with sustained musical instruments seems less intuitive than using percussive sounds such as piano notes or plucked violin strings. The temporal evolution of the features was not considered in this work. Sounds that vary in time are expected to pose a greater challenge to orchestrate using notes of musical instrument sounds. Most musical instruments from an orchestra can be played with temporal variations, such as *glissando* or *vibrato*. Thus it seems natural to use target sounds that vary in time.

Future work should investigate how to incorporate time in the search for orchestrations to find better combinations for target sounds that present temporal evolution. The attack time is an important feature to distinguish percussive from sustained sounds. Target sounds with a high degree of temporal variation such as the *Elephant* require a fitness function that encodes temporal variations of the features. Finally, the assumption that orchestrations that are more similar to the target sound are aesthetically better could be investigated.

The results obtained in this work were made available online in a dedicated webpage[3].

Acknowledgments. This work is financed by the FCT - Fundação para a Ciência e a Tecnologia (Portuguese Foundation for Science and Technology) within project "UID/EEA/50014/2013." The authors would like to thank the integrated masters program in Electrical and Computer Engineering (MIEEC) from the University of Porto (FEUP) for the financial support.

References

1. Caclin, A., McAdams, S., Smith, B., Winsberg, S.: Acoustic correlates of timbre space dimensions: a confirmatory study using synthetic tones. J. Acoust. Soc. Am. **118**(1), 471–482 (2005)
2. Camacho, A., Harris, J.: A sawtooth waveform inspired pitch estimator for speech and music. J. Acoust. Soc. Am. **124**(3), 1638–1652 (2008)
3. Carpentier, G., Tardieu, D., Assayag, G., Rodet, X., Saint-James, E.: Imitative and generative orchestrations using pre-analysed sound databases. In: Proceedings of the Sound and Music Computing Conference, pp. 115–122 (2006)
4. Carpentier, G., Tardieu, D., Assayag, G., Rodet, X., Saint-James, E.: An evolutionary approach to computer-aided orchestration. In: Giacobini, M. (ed.) EvoWorkshops 2007. LNCS, vol. 4448, pp. 488–497. Springer, Heidelberg (2007)

---

[3] Access http://goo.gl/4l9NqX to listen to the target sounds and results.

5. Carpentier, G., Assayag, G., Saint-James, E.: Solving the musical orchestration problem using multiobjective constrained optimization with a genetic local search approach. J. Heuristics **16**(5), 681–714 (2010)
6. Carpentier, G., Tardieu, D., Harvey, J., Assayag, G., Saint-James, E.: Predicting timbre features of instrument sound combinations: application to automatic orchestration. J. New Music Res. **39**(1), 47–61 (2010)
7. de Castro, L., Timmis, J.: An artificial immune network for multimodal function optimization. In: CEC 2002, Proceedings of the 2002 Congress on Evolutionary Computation, vol. 1, pp. 699–704, May 2002
8. Goto, M., Hashiguchi, H., Nishimura, T., Oka, R.: RWC music database: popular, classical and Jazz music databases. In: Proceedings of the International Society for Music Information Retrieval Conference, vol. 2, pp. 287–288 (2002)
9. Grey, J.: Multidimensional perceptual scaling of musical timbres. J. Acoust. Soc. Am. **61**(5), 1270–1277 (1977)
10. Handelman, E., Sigler, A., Donna, D.: Automatic orchestration for automatic composition. In: 1st International Workshop on Musical Metacreation (MUME 2012), pp. 43–48. AAAI (2012)
11. Hummel, T.: Simulation of human voice timbre by orchestration of acoustic music instruments. In: Proceedings of the International Computer Music Conference (ICMC), p. 185 (2005)
12. Kendall, R.A., Carterette, E.C.: Identification and blend of timbres as a basis for orchestration. Contemp. Music Rev. **9**(1–2), 51–67 (1993)
13. Krumhansl, C.L.: Why is musical timbre so hard to understand? Struct. Percept. Electroacoust. Sound Music **9**, 43–53 (1989)
14. Lartillot, O., Toiviainen, P.: A matlab toolbox for musical feature extraction from audio. In: International Conference on Digital Audio Effects, pp. 237–244 (2007)
15. McAdams, S., Giordano, B.L.: The perception of musical timbre. In: Hallam, S., Cross, I., Thaut, M. (eds.) The Oxford Handbook of Music Psychology, pp. 72–80. Oxford University Press, New York (2009)
16. McAdams, S., Winsberg, S., Donnadieu, S., De Soete, G., Krimphoff, J.: Perceptual scaling of synthesized musical timbres: common dimensions, specificities, and latent subject classes. Psychol. Res. **58**(3), 177–192 (1995)
17. Psenicka, D.: SPORCH: an algorithm for orchestration based on spectral analyses of recorded sounds. In: Proceedings of International Computer Music Conference (ICMC), p. 184 (2003)
18. Rose, F., Hetrik, J.E.: Enhancing orchestration technique via spectrally based linear algebra methods. Comput. Music J. **33**(1), 32–41 (2009)
19. Tardieu, D., Rodet, X.: An instrument timbre model for computer aided orchestration. In: 2007 IEEE Workshop on Applications of Signal Processing to Audio and Acoustics, pp. 347–350. IEEE (2007)

Evolving Atomic Aesthetics and Dynamics

Edward Davies[1], Phillip Tew[2], David Glowacki[3], Jim Smith[1],
and Thomas Mitchell[1]($\boxtimes$)

[1] University of the West of England, Bristol, UK
tom.mitchell@uwe.ac.uk
[2] Interactive Scientific, Bristol, UK
[3] University of Bristol, Bristol, UK

Abstract. The depiction of atoms and molecules in scientific literature owes as much to the creative imagination of scientists as it does
to scientific theory and experimentation. danceroom Spectroscopy (dS)
is an interactive art/science project that explores this aesthetic dimension of scientific imagery, presenting a rigorous atomic simulation as an
immersive and interactive installation. This paper introduces new methods based on interactive evolutionary computation which allow users -
both individually and collaboratively - to explore the design space of dS
and construct aesthetically engaging visual states. Pilot studies are presented in which the feasibility of this evolutionary approach is discussed
and compared with the standard interface to the dS system. Still images
of the resulting visual states are also included.

Keywords: Aesthetic evolution · Evolutionary computation · Interactive evolutionary computation · Molecular aesthetics

1 Introduction

Atoms and molecules are too small for us to see and consequently their detailed
and often colourful depiction in scientific literature originates as much from the
imagination of scientists as it does from reality [1]. This 'molecular aesthetic' may
be intended primarily to convey information and stimulate new understanding,
but it also reveals a hidden beauty in the natural world which has inspired new
forms of contemporary art [2] and music [3].

Interactive evolutionary computation (IEC) has been used across a range of
artistic domains to explore complex design spaces and support artists in the
production of novel work [4]. In this paper the potential for IEC is explored
for evolving the aesthetics and dynamics of an interactive atomic simulation
engine called danceroom Spectroscopy (dS). Following a short description of
the dS system and IEC, two interfaces supporting single- and collaborative-user
evolution are introduced and studied. Examples of the imagery produced within
these studies are provided and discussed, followed by concluding remarks and
plans for future work.

© Springer International Publishing Switzerland 2016
C. Johnson et al. (Eds.): EvoMUSART 2016, LNCS 9596, pp. 17–30, 2016.
DOI: 10.1007/978-3-319-31008-4_2

2   danceroom Spectroscopy

danceroom Spectroscopy is an internationally acclaimed art/science project fusing interactive technology with rigorous particle physics to create an interactive and immersive simulation of the atomic nanoworld, see Fig. 1. Multiple participants are able to simultaneously traverse and interact with the system through whole body interaction as their image is captured using an array of depth cameras and embedded into an atomic simulation as electrostatic fields. dS has been deployed as the framework underpinning a series of international dance/music performances, art/science/technology installations, and as an educational platform for teaching science and dance to school-age children. For fuller background, mathematical, technical and artistic expositions see [5,6].

The dS physics and graphical rendering engine is highly flexible, capable of simulating extremely diverse visual and interactive states. It is manually controlled using a dedicated graphical user interface (GUI), shown in Fig. 4. Due to the complexity of the design space, configuring a dS state requires that users possess expert knowledge of atomic physics and computer graphics. Consequently, this interface is rarely exposed to general participants, whose experiences of the system are normally limited to interacting with a number of pre-programmed states. In dance, educational, performance and installation scenarios, the GUI to dS has been observed to present novice users and collaborators with a significant obstacle when attempting to explore the capabilities of the system in an improvisatory manner. Typically, they tend to work closely with an experienced user,

Fig. 1. danceroom Spectroscopy

verbally describing their intentions - a process that is highly inefficient. However, this barrier to entry can be elegantly addressed with the development of an IEC interface to dS, enabling the domain specific controls of the simulation to be abstracted by a simpler process, guided by the subjective aesthetic preferences of a user.

3   Interactive Evolutionary Computation (IEC)

Interactive evolutionary computation is a long-established and powerful optimisation method for finding solutions to complex design problems. IEC has been utilised across a range of disciplines, most notably in art and design but also within domains such as science and engineering [7]. IEC's synthesis of computational optimisation and instinctive human enquiry has been shown to produce higher-quality solutions more rapidly than when humans and machines work independently [8–10]. One of the principle advantages of IEC over more traditional optimisation methods is that it enables design solutions to be evaluated by a human user, without the need for a formalised objective function, a requirement of canonical evolutionary algorithms. As the dS design space is large and already requires the subjective evaluation of visual and interactive features, IEC offers the potential for a powerful alternative to the existing GUI interface. The application of IEC to dS also opens up a novel opportunity to explore the interactive evolution of atomic aesthetics.

4   Evolving Atomic Aesthetics

At the heart of dS is a rigorous atomic dynamics simulation based upon Hamilton's equations of motion, which is a common approach used to study the dynamics of classical and quantum molecular systems. The Cartesian coordinates of up to 30,000 simulated atoms are computed using dedicated GPU accelerated software on a high-performance workstation at 60 frames per second [5]. The system is configured via a software GUI, which is shown in Fig. 4.

To enable the dS visual and interactive design space to be explored using IEC, an interactive evolutionary algorithm was created based upon the $(1 \dotplus \lambda)$ Evolution Strategy (ES) [11]. In this approach, a population of individuals is evaluated at each generation from which a single parent is selected and mutated to produce λ offspring. For ease of development and experimentation, this algorithm and its interface were developed as a separate application to the main dS system, communicating through an existing Open Sound Control [12] interface.

4.1   Representation

A subset of 60 dS parameters control the atomic dynamics and aesthetic representation of the system. These include physics parameters (such as atom size, simulation temperature and number of atoms) as well as visual parameters (including atom/electrostatic field colour, time-history visibility and colour inversion). Consequently, a vector of 60 values for each of these parameters represents a single solution within the dS design space and the genome of a single individual within the IEC population. The majority of these parameters take real valued numbers in the range $0 \rightarrow 1.0$; however, a minority of parameters have binary state values, represented by a boolean in the solution vector.

4.2 Evaluation and Selection

To assess an individual for fitness, the unique combination of parameters comprising its genotype is transferred to dS where it is expressed as a visual and interactive state (or *phenotype*) whose aesthetic appeal may be evaluated by a human observer [13]. The computational complexity and presentation format of dS limits users to evaluating one individual at a time. Consequently, the IEC interfaces described later were developed to enable users to evaluate each population member in sequence before selecting a single individual to act as a parent of the subsequent generation [7]. The fitness of a dS state is determined by its aesthetic and interactive merit (i.e. the extent to which the user likes and finds it engaging). As part of the selection process the user must provide a fitness score in the range $0 \rightarrow 1.0$, enabling the user to steer the process towards increasingly optimised regions of the search space.

4.3 Mutation

When the user selects a parent individual from the population, the next generation of progeny is created exclusively by mutation. The mutation process in this work is also taken from the standard ES algorithm: real valued parameters from the parent individual are copied to each offspring with the addition of an independent random number drawn from a standard normal distribution. Binary values are flipped when the outcome of a simulated dice roll is 1. In both cases, the mutation strength is linked with the fitness value the user ascribes to the parent, such that on average larger mutations are observed when parents are awarded lower fitness to promote exploration. Conversely, smaller mutations occur, when parents are of high fitness to promote optimisation. In practice this mutation step size adjustment is achieved for the real valued parameters by setting the standard deviation (σ) of the random normal mutation inversely proportional to an individual's fitness (f): $\sigma = (1 - f)^2$. For binary values the number of sides on the 'die' is set inversely proportional to the individual's fitness: $s = \lceil \frac{1}{f^2} - 1 \rceil$, with no flip performed when $f = 0$. Consequently, parents assigned a fitness value of 0.0 will undergo very strong mutation and parents with fitness values of 1.0 will not be mutated.

5 Experimentation

Two distinct and practical use cases were considered in which users could evolve atomic aesthetics with IEC. First, single-user scenarios where a user creates atomic visual states rapidly in isolation in preparation for dance performances and art installations. Secondly, collaborative-user scenarios in which groups could collectively contribute to the design of a visual state. For example, this might take place during installations or performances where visitors or audience members are able to form a consensus to help evolve designs over longer time periods.

5.1 Single-User IEC Interface

The single-user IEC interface, shown in Fig. 2, is written in C++ using the JUCE application framework [14] and modelled on the seminal IEC example: Dawkins' 'Blind Watchmaker' program for evolving *biomorph* creatures [15]. The interface presents users with a grid of nine buttons, each representing an individual member of the population. By hovering their mouse over a grid button, users are able to load the associated individual into the running dS engine where it is rendered instantaneously onto a separate display. Users are able to rapidly switch between the nine population members while remaining immersed in the dS experience. When ready, the user can click the button associated with their preferred individual to select it as the parent of the next generation. The underlying evolutionary algorithm implements a $(1 + 8)$ *elitist* strategy, meaning that each new generation comprises the selected parent along with eight mutants. Users continue this evolutionary cycle until either a satisfying solution is located or they become fatigued.

Following an informal heuristic evaluation with the project contributors [16], subtle modifications to the interface were made to enhance user experience.

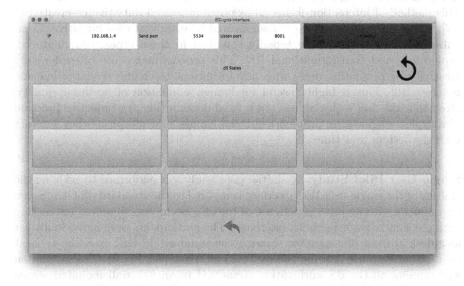

Fig. 2. Single-User IEC Interface

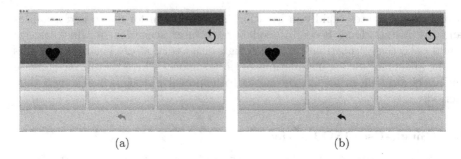

(a)   (b)

Fig. 3. Single-User IEC interface (a) auditioning a candidate solution and (b) candidate selected as parent

For example, when candidate solutions are being auditioned, their associated button turns red and a heart image appears at its centre (see Fig. 3a). The size of the heart varies with the x-position of the mouse cursor, enabling the user to express the degree to which they like the associated individual. Upon selection with a single click, the size of the heart is converted to a fitness value, assigned to the chosen individual and factored into the subsequent mutation process as described in Sect. 4.3. At the turn of each generation, the selected individual persists to the next generation: its button turns green (see Fig. 3b) and the remaining individuals are replaced with eight mutated offspring.

The immediacy of this IEC interface allows the user to create and review a variety dS states many times more rapidly than would otherwise be possible using the standard GUI. The process is also unbiased in its exploration, jumping to regions of the design space that might not typically be encountered manually. This method of interaction forms an example of 'mixed initiative co-creativity' [17], where the user's creativity is augmented by computational initiative. Computational creativity can be broadly classified in three types: combinatorial, exploratory and transformational [18]. This process makes use of exploratory creativity by presenting users with novel solutions to enable *serendipitous discoveries* [10]. This is highly useful for human evaluation of aesthetic quality, given that the user's objective is not a fixed domain (like the objective function of a typical engineering problem) but can change and morph as they interact with the system over time.

Single-User Pilot Study. An initial pilot study was performed to evaluate the efficacy of the single-user IEC interface compared to the standard dS GUI. Seven participants with no prior experience of the dS system were observed interacting with the simulation using both interfaces. The participants were individually sat at a desk in front of a monitor showing the standard dS GUI (see Fig. 4), next to a second monitor showing the IEC grid interface (see Fig. 3). The simulation was running on the dS workstation connected to an 83" wall mounted screen positioned in front of the user.

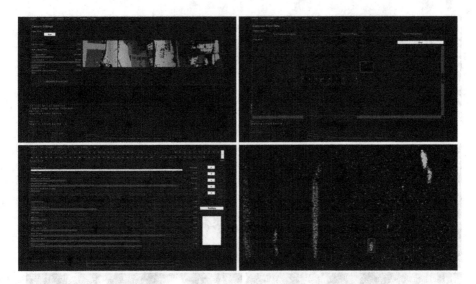

Fig. 4. dS GUI

Each user was given a maximum of four minutes to design an aesthetically engaging visual and interactive state for dS using each interface type. The first interface encountered was alternated for each participant in an attempt to control any effects that ordering might have on the results. Participants were subsequently asked to comment on their experience with each interface and to answer a series of questions. The results are summarised in Table 1 below.

Table 1. Summary of single-user IEC interface pilot study

Profession	Time (dS/IEC)	Preferred preset	Preferred interface
Lecturer	4:00/3:30	IEC	dS
Theatre director	4:00/2:00	IEC	dS
Producer	4:00/2:30	IEC	IEC
Research manager	4:00/2:30	IEC	IEC
Graphic designer	3:30/1:50	IEC	dS
Student/intern	4:00/4:00	IEC	dS
Studio assistant	4:00/4:00	IEC	IEC

Figure 5 shows screenshots of the dS states created by four participants during the pilot study; the left hand panel shows states designed using the standard dS GUI and the right hand panel shows states designed using the IEC interface:

The states designed using the IEC interface are much more vibrant and exhibit greater variation than those made manually using the dS GUI, and this

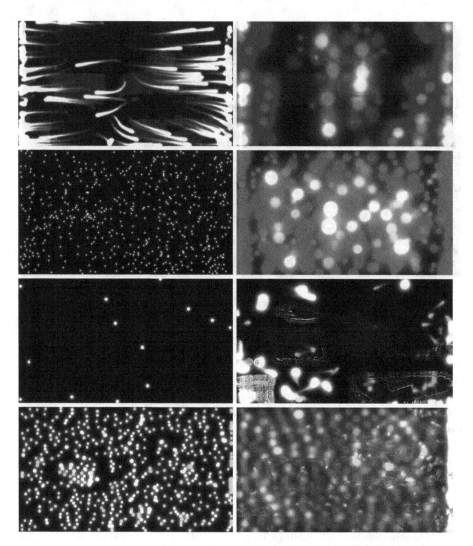

Fig. 5. Example dS states created by four participants; left images designed using the dS GUI, right images designed using the IEC interface

is reaffirmed by the fact that all seven participants stated that they preferred their evolved state. Many participants reported feeling overwhelmed by the sheer number of parameters and their complex names in the dS GUI and few managed to explore beyond the first ten within the time limit. States developed using the standard dS interface exhibit very little aesthetic variance and are all clustered around the experimental starting point within the design space. In four minutes participants were unable to explore the full range of aesthetic potential, and many design aspects including colour remained unchanged. For example, only one person managed to find the parameter controlling the visibility of their electrostatic avatar. However, all of these parameters were simultaneously modified by the IEC interface, allowing participants to traverse a wide range of aesthetic states much more freely. This is reinforced by the differences in time spent using each interface: six out of seven participants ran out of time when using the dS interface. On the other hand, most participants were able to reach a satisfying state within the same time limit using the IEC interface.

Participants also stated that they felt more engaged with the simulation using the IEC interface as they were not focused on the intricacies of the GUI, instead devoting their attention to the on-screen aesthetics of dS. However, it is clear from the results that the majority of participants preferred to use the standard dS interface over the IEC. The main reason given for this was that participants felt they had more direct control over the simulation and were able to more precisely and intentionally change aspects of the visualisation. The IEC interface was often described as more "random" and "chaotic". However, one participant particularly enjoyed this stochastic behaviour, comparing the differences between the interfaces to selecting music: "the joy of shuffling rather than choosing a song that you like". Most participants preferring the dS GUI did note that it would take a great deal of time to learn to control the system effectively and that the IEC interface presents a much lower barrier to entry.

One feature that was utilised extensively throughout the study by participants was the undo function. Many participants realised that they had reached a "dead end" and chose to revert to a previous generation. This feature was noted to enhance the evolution of dS states as participants were able to revert back to earlier positions in the search space and explore an alternative design path.

5.2   Collaborative IEC Interface

As the previous study indicates, IEC is an excellent tool for enabling single-users to explore large, subjective fitness landscapes. A second scenario that was considered allowed multiple users to collaboratively control the IEC process, evaluating and selecting individuals by majority consensus. This method of crowdsourced evaluation has been adopted in a number of previous IEC studies to create collaboratively evolved music [19] and imagery [8]. Collaborative IEC methods have the advantage of leveraging the efforts of multiple evaluators to address problems of user fatigue [20], enabling prolonged evolution such that fitter and more universally engaging designs can emerge. Given that dS is frequently deployed as an art installation for extended periods, a collaborative approach is more likely to result in wider exploration of the design space and ensures that the experience is continually changing as participants contribute to the evolutionary process.

To support the collaborative evolution of dS states, the underlying algorithm was modified to a $(1 + 1)$ strategy. That is, at each generation the parent is mutated to produce a new offspring, which is loaded into the dS system for evaluators to observe. If the new offspring receives a positive reaction it replaces its parent. However, if the new individual receives a negative reaction, it is discarded and the original parent persists.

Fig. 6. Collaborative IEC interface web app running on a smartphone device

An interface to the collaborative IEC process was developed to accommodate any number of users participating using their smartphones, joining or leaving a voting network as they enter and leave the installation. A simple web application was designed and developed, enabling participants to either 'like' or 'dislike' the currently loaded dS state. The main program running the evolutionary algorithm was again written in C++ using JUCE and the voting service was written in JavaScript using node.js [21]. Users connect to an open Wi-Fi network and browse to the IP address of the host machine, which serves a simple web page featuring 'like', 'dislike' buttons and labels displaying the current votes of the group (number of likes/dislikes). A countdown timer is also shown to indicate when the next state will be loaded (see Fig. 6). The duration of the countdown is variable and can be easily adjusted to suit the environment. For example, in the pilot study described in the next section, a countdown of 20 seconds was used to maintain the focus of the audience and make the most of the time available. However, in a more permanent context, such as an art installation, a longer countdown would be more appropriate, providing time for participants to interact and play with states between transitions. Connected users are able to 'like' or 'dislike' the current state and when the countdown reaches zero, recorded votes are converted to a fitness value, which is used to mutate the selected parent as described in Sect. 4.3. Fitness at the end of each generation was calculated as the normalised ratio of the number of likes (l) to dislikes (d): $\frac{l/d}{1+(l/d)}$.

Collaborative IEC Pilot Study. To assess the potential of the collaborative IEC interface, a short pilot study was conducted during an undergraduate Music Technology lecture at the University of the West of England with 48 participants. The dS simulation was projected onto a large screen at the front of

Fig. 7. Collaborative IEC pilot study

the lecture theatre and participants were invited connect to the web server using their smartphones, tablets and/or laptops and to observe or interact with the simulation as shown in Fig. 7.

The session began with a brief introduction to dS, a description of the experiment and instructions on how to participate. The system was then set running for 25 min, during which time 75 generations elapsed. A range of states were created in this time; a screenshot of one example is shown in Fig. 8. Following the experiment, a short group discussion was conducted to gather feedback.

The participants commented that they enjoyed the process and liked many of dS visual states that emerged during the experiment. However, there was not always a clear connection between user votes and the subsequent response from the system. A positive consensus would not always produce changes that participants perceived to be positive. In these circumstances a strong positive consensus was frequently followed by a strong negative consensus, if the connecting mutation was judged to be detrimental. From a user perspective there appears to be an expectation that positive votes should produce positive changes; however, an unbiased mutation operator guarantees that offspring will not always be fitter than their parents. This factor could have been improved with a clearer description of the underlying evolutionary process at the beginning of the experiment to help manage user expectations.

Another point raised for discussion was the inability of the web application GUI to capture the nuances motivating participants to vote. For example,

Fig. 8. Example state produced during the collaborative IEC pilot study

participants frequently used the 'dislike' button when they became bored with the current state and not necessarily because they disliked it. Perhaps improvements to this interface might incorporate a 'change' button to disambiguate fatigue from a genuine dislike. Additionally, more elaborate parallel IEC algorithms might be used in which multiple states are evolved in parallel, and transitions between these different states might help to alleviate fatigue.

6   Conclusion

In this paper, interactive evolutionary computation (IEC) has been used to evolve the aesthetics and dynamics of an interactive and immersive atomic simulation called danceroom Spectroscopy (dS). Two interfaces to the IEC process were presented that enable single- and collaborative-user evolution of parameters to support the artistic process of visual state design. Both interfaces were then used within pilot studies to assess their utility within these contexts.

The single-user IEC interface was studied with seven participants comparing the manual GUI interface with the IEC interface. While the IEC interface was shown to make it simple to rapidly explore the design space, direct and manual control of individual parameters was highlighted as an important feature of the dS GUI. This study suggests that it may be useful in future to consider additional forms of user interaction. For example, in addition to the metaheuristic global search, users could select certain aspects of the design space to evolve, freezing the remaining parameters; a method which has been shown previously to increase the quality and rate of search [22]. In application to dS, this might involve freezing all parameters except those affecting the characteristics of the atoms, or other distinct features of the simulation. In practice, a combination of both interfaces could support the creative process for experts, using the IEC interface to explore the design space before switching to the manual interface for fine-tuning parameters.

The collaborative-user IEC interface was developed to give any number of dS performance and installation attendees the ability to contribute to the design of visual states using their smartphone devices. The study successfully produced a broad range of aesthetic states and subsequent discussion revealed a perceived separation between individual votes and the subsequent response by the system. This was due to the diluting effects of aggregating votes to form a consensus, but also a feature of the mutation operator. The collaborative interface could certainly be developed further if it were deployed for an extended period within the context of an art installation. For example, alternative evolution strategies incorporating recombination could be explored, seeded with existing states that have been developed manually during previous installations. Each state could be evolved in parallel to ensure that users experience a variety of high-quality visual states which may help to reduce user fatigue and increase interaction times. Evaluation might also be gathered from alternative sources. For example, given that the movements of participants are already captured by dS, these movements could be classified to gauge the reactions of participants. Fitness evaluation may then take place without users having to vote explicitly. These avenues of research will certainly form the basis of future work and along with the techniques proposed in this paper, could have the potential for application within a wider range of digital arts platforms with large design spaces.

References

1. Flannery, M.C.: Goethe and the molecular aesthetic. Janus Head **8**(1), 273–289 (2005)
2. Voss-Andreae, J.: Quantum sculpture: art inspired by the deeper nature of reality. Leonardo **44**(1), 14–20 (2011)
3. Fernández, J.D., Vico, F.: AI methods in algorithmic composition: a comprehensive survey. J. Artif. Intell. Res. **48**(1), 513–582 (2013)
4. Lewis, M.: Evolutionary visual art and design. In: Romero, J., Machado, P. (eds.) The Art of Artificial Evolution. Natural Computing Series, pp. 3–37. Springer, Berlin Heidelberg (2008)
5. Glowacki, D.R., O'Connor, M., Calabro, G., Price, J., Tew, P., Mitchell, T.J., Hyde, J., Tew, D., Coughtrie, D.J., McIntosh-Smith, S.: A GPU-accelerated immersive audiovisual framework for interaction with molecular dynamics using consumer depth sensors. Faraday Discuss. **169**, 63–87 (2014)
6. Mitchell, T., Hyde, J., Tew, P., Glowacki, D.: Danceroom Spectroscopy: At the frontiers of physics, performance, interactive art and technology, Leonardo **49**(2), (2016)
7. Takagi, H.: Interactive evolutionary computation: fusion of the capabilities of EC optimization and human evaluation. Proc. IEEE **89**(9), 1275–1296 (2001)
8. Secretan, J., Beato, N., Ambrosio, D.B.D., Rodriguez, A., Campbell, A., Folsom-Kovarik, J.T., Stanley, K.O.: Picbreeder: a case study in collaborative evolutionary exploration of design space. Evol. Comput. **19**(3), 373–403 (2011)

9. do Nascimento, H.A.D., Eades, P.: User hints: a framework for interactive optimization. Future Gener. Comput. Syst. **21**(7), 1177–1191 (2005)
10. Woolley, B.G., Stanley, K.O.: A novel human-computer collaboration: combining novelty search with interactive evolution. In: Proceedings of the 2014 Conference on Genetic and Evolutionary Computation, NY, USA, 233–240. ACM, New York (2014)
11. Beyer, H.G., Schwefel, H.P.: Evolution strategies - a comprehensive introduction. Natural Comput. **1**(1), 3–52 (2002)
12. Wright, M., Freed, A., Momeni, A.: Opensound control: state of the art 2003. In: Proceedings of the 2003 Conference on New Interfaces for Musical Expression (NIME 2003), pp. 153–160. National University of Singapore, Singapore (2003)
13. Serag, A., Ono, S., Nakayama, S.: Using interactive evolutionary computation to generate creative building designs. Artif. Life Robot. **13**(1), 246–250 (2008)
14. Storer, J.: JUCE (Jules' Utility Class Extensions) (2015). http://www.juce.com. Accessed 25 October 2015
15. Dawkins, R.: The Blind Watchmaker: Why the Evidence of Evolution Reveals a Universe without Design. WW Norton and Company, New York (1986)
16. Nielsen, J., Molich, R.: Heuristic evaluation of user interfaces. In: Proceedings of the ACM CHI 1990 Conference, pp. 249–256. Seattle, WA (1990)
17. Yannakakis, G.N., Liapis, A., Alexopoulos, C.: Mixed-initiative co-creativity. In: Proceedings of the 9th Conference on the Foundations of Digital Games (2014)
18. Boden, M.A.: The Creative Mind: Myths and Mechanisms. Psychology Press, New York (2004)
19. MacCallum, R.M., Mauch, M., Burt, A., Leroi, A.M.: Evolution of music by public choice. Proc. Natl. Acad. Sci. **109**(30), 12081–12086 (2012)
20. Lee, J.Y., Cho, S.B.: Sparse fitness evaluation for reducing user burden in interactive genetic algorithm. In: Fuzzy Systems Conference Proceedings, vol. 2, pp. 998–1003. IEEE (1999)
21. Joyent, Inc. (2015) Node.js homepage. http://nodejs.org/. Accessed 10 October 2015
22. Pauplin, O., Caleb-Solly, P., Smith, J.E.: User-centric image segmentation using an interactive parameter adaptation tool. Pattern Recogn. **43**(2), 519–529 (2010)

Augmenting Live Coding with Evolved Patterns

Simon Hickinbotham$^{(\boxtimes)}$ and Susan Stepney

YCCSA, University of York, York, UK
sjh518@york.ac.uk

Abstract. We present a new system for integrating evolvutionary processes with live coding. The system is built upon an existing platform called Extramuros, which facilitates network-based collaboration on live coding performances. Our evolutionary approach uses the Tidal live coding language within this platform. The system uses a grammar to parse code patterns and create random mutations that conform to the grammar, thus guaranteeing that the resulting pattern has the correct syntax. With these mutations available, we provide a facility to integrate them during a live performance. To achieve this, we added controls to the Extramuros web client that allows coders to select patterns for submission to the Tidal interpreter. The fitness of the pattern is updated implicitly by the way the coder uses the patterns. In this way, appropriate patterns are continuously generated and selected for throughout a performance. We present examples of performances, and discuss the utility of this approach in live coding music.

1 Introduction

We explore the use of evolutionary techniques to generate new musical constructs during live coding performances. Live coding is the use of domain-specific languages (DSLs) to improvise new musical pieces in a live concert setting. We present a new evolutionary system which augments this process by generating and maintaining a population of coded musical patterns that are interactively expressed as audio during the performance. The system allows the actions of the human coders to be fed back to the population as adjustments to the fitness of patterns, resulting in a novel musical interactive genetic algorithm (MIGA, [1]). The system is sufficiently flexible to allow the generation of pieces using evolved patterns alone, or any mix of evolved patterns and manually configured patterns.

Using artificial evolution to generate music has a long history. The core approach is usually to generate populations of complete musical pieces and to assign fitness values to each individual piece via human perception [1]. This process requires that each generated piece is heard at least once, and so causes a significant bottleneck in the evolutionary process. In addition, there is the issue of *listener fatigue*, in which the human assignment of fitness values varies as the listener wearies of listening to large numbers of generated pieces. There are ways around this problem, usually by either only presenting a subset of the generated pieces to the listener [2], or making the pieces very short [3]. Both of these

© Springer International Publishing Switzerland 2016
C. Johnson et al. (Eds.): EvoMUSART 2016, LNCS 9596, pp. 31–46, 2016.
DOI: 10.1007/978-3-319-31008-4_3

solutions have obvious drawbacks both in terms of the evaluation of each piece and the eventual result. In addition, the *positive* side of changes in the human evaluation of a musical pattern throughout a performance has received relatively little attention in previous work. Recognising when a repeating pattern should be changed is an important part of the composition process.

Our approach differs from the above methodologies as follows. Firstly, we are using the paradigm of *live coding* [4] as our musical framework. Here, rather than evolve entire pieces of music *ab initio*, the evolutionary process is interactive, and takes place while the piece is being performed. This allows for a more natural interaction between the evolutionary mechanism and the composer/performer. The recent development of live coding systems affords a new opportunity for researchers in evolving systems because new musical patterns must be generated many times during a performance. We give the artist (here referred to as the *live coder* or just *coder*) a new facility to augment their hand-designed patterns with ones that are generated by the evolutionary algorithm, and to blend the two together as they see fit. The evolutionary system *augments* the process of improvisation by generating novel patterns which the live coder can integrate and respond to. The population of evolved patterns changes gradually with the piece, and provides a reference point and storage for a changing bank of aesthetically pleasing patterns.

The recent development of the Extramuros [5] system has made the implementation of this idea more generic, and easier to achieve. Extramuros is a system that allows browser-based collaboration of a group of performers on live-coded pieces of music and graphics across a network. The availability of this system makes it easier to use automation to suggest patterns for use and further manipulation by the coder. An evolutionary process is an ideal candidate for the automatic generation of new coding patterns within this context.

Tidal [4] is the live coding language that we use in this work. Other live coding languages exist, such as Sonic Pi [6] and Chuck [7]. We selected Tidal as our live coding language because it is relatively well documented and straightforward to install in an Extramuros framework. The live coding approach allows more immediate interaction between human and automaton. Live coding languages are more like conventional computer code than other musical notation systems but with more emphasis on brevity and ease of editing than is common in programming languages, sometimes at the expense of clarity. Live coding requires a text-based grammar to encode the musical sounds, designed to be constantly manipulated rather than simply written once and interpreted many times. This grammar is amenable to a genetic programming (GP) approach to evolving systems.

The experimental work we present here tests this idea as a proof of concept. We have augmented the Extramuros system with an interactive evolutionary algorithm, which maintains a population of patterns on the client, and uses a separate server to carry out the process of parsing and mutating individuals in the population.

Evolutionary systems require a mechanism for mutation. A naive mutation model would substitute a single ASCII character in the pattern for another, but this would rarely yield a pattern that could be parsed by the interpreter. Clearly, a more sophisticated mutation scheme is required which preserves the syntax of the coded pattern. One approach would be Genetic Programming (GP), in which a tree representation of a program is used to organise the options of mutation. However, GP still has limitations, because it must be possible to interchange any node on the tree representation.

2   Methodology

We first describe the workflow of a performance, as this will give a clear understanding of the complete system and how it is used in a live performance setting. Following from this, we present a more detailed description of the evolutionary aspects of the system that we have implemented to test the concept.

Our evolutionary system is built around the Extramuros platform, Fig. 1. In a standard Extramuros system, each performer controls their contribution to the performance by entering Tidal code into text-boxes in a web-browser-based client (see https://www.youtube.com/watch?v=zLR02FQDqOM). The client can be configured to have any number of input boxes, and each user writes their code in one or more of them. The Extramuros server uses ShareJS, a library for concurrently editing content over a network. This means that each performer can see the code that every other performer in the ensemble is creating in near-real time in every browser. Indeed, it is only by agreed convention between performers that they do not attempt to edit each other's code, although there are no barriers to doing so.

When a performer has completed the latest edit to their code, they submit it to the Tidal interpreter by pressing the 'eval' button next to the text box, or by using a keyboard shortcut. (There were versions of Tidal which did not need this 'submission' step, but it was found that this led to jarring audio as the new pattern was being typed in.) A piece of edited code is called a *pattern*. The Tidal interpreter parses the pattern and sends the resulting code to a synthesiser, usually the Dirt synthesiser that Tidal was developed to control. Other compatible

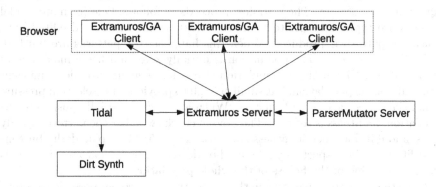

Fig. 1. Overview of the system

synthesisers can also be used, and recent developments in Tidal also allow it to send MIDI data.

In order to allow evolution of patterns, we developed a client-side genetic algorithm and an additional server, shown in Fig. 1 as the ParserMutator server, which checks the syntax of each pattern as it is added the population, and rejects patterns which do not conform to the grammar. When requested, the ParserMutator server also generates mutated patterns which are syntactically correct. A population of patterns is held in the Document Object Model (DOM) of each client page. Each user can maintain their own population of patterns in this way without interfering with other coders in the performance, although a coder can push code that appears in another coder's text box to their own population if they want to. The core idea is that aesthetically pleasing patterns are 'pushed' to the GA population, and mutants of these patterns are 'pulled' from it later on. The fitness of a pattern is increased or decreased according to the way it is used in the performance. Since the system is designed to operate in a live performance, it is important to let the performer(s) decide when to do this. To facilitate this flexibility, we added five new buttons to each text-entry box, giving a total of six operations:

1. **eval:** Send the pattern to the Tidal interpreter
2. **push:** Send the pattern to the population (via the ParserMutator)
3. **pull:** Pull a pattern from the population
4. **pullmut:** Request a mutation of the current pattern from the ParserMutator
5. **up:** Increase the fitness of the current pattern (if it exists in the population)
6. **dn:** Decrease the fitness of the current pattern; remove it from the population if fitness < 0

We describe how these operations are used to change the fitness of a pattern below (Sect. 2.3). In addition to the extra controls, we added a small notifications area to each box so that the client can communicate to the user the effect of the last action in each box. For the layout of a each text entry box, see the lower panel of Fig. 2.

The actions that these buttons encapsulate are connected into a performance workflow as shown in Fig. 3. The commands for these actions were written in JQuery and JavaScript. The workflow proceeds as follows. One (or more) Tidal patterns are typed into the text entry box(es) by the coder, and edited until a pleasing pattern is produced. When this happens, the patterns are 'pushed' to the GA population. The population is initially zero, and has a maximum of 20 individuals. This limit on population size is large enough to allow sufficient variability in the population, but small enough to provide some selection pressure within the timescale of a performance. When a pattern is initially pushed to the population, its fitness is set to 1. The system checks whether the pattern already exists, and if it does then its fitness is increased by 0.5. The 'up' and 'dn' buttons allow fitness to be respectively increased or decreased by 0.25 at any time. This allows fine-tuning of the fitness of the whole population.

Once there are some patterns in the population, it becomes possible to copy a pattern from the population into a text box using the 'pull' button. An individual

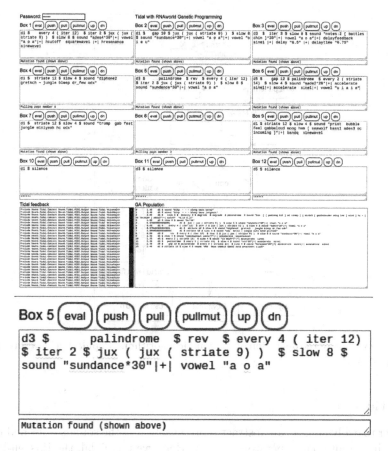

Fig. 2. The evolving Extramuros platform (top), and closeup view of pattern entry box 5 (bottom) (colours inverted for clarity) (Color figure online).

pattern is selected randomly from the population, with the chance of being selected weighted by the fitness of each individual. One copied to a text box, the pattern can be submitted to the Tidal interpreter, edited manually or mutated by pressing the 'pullmut' button.

When the maximum permitted number of patterns are in a population, new patterns that are submitted to the population take the place of the least fit existing pattern. If there is more than one pattern with the minimum fitness value, then one of these low-scoring patterns is selected for deletion at random.

Having described the overall function of the system, we now turn to the mechanism by which we maintain a population of Tidal patterns, and how we are able to mutate them. This is achieved by reference to a grammar, from which we construct a pattern parser and a mutation operator to build our genetic algorithm (GA). These are described below, after a brief description of the Tidal pattern language.

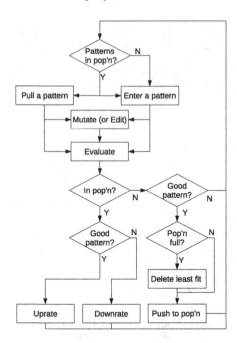

Fig. 3. Flowchart of the operation of the genetic algorithm. "Population" is abbreviated to "pop'n"

2.1   Tidal

In order to evolve music, an abstract representation of the sounds is needed upon which the evolutionary operators can act. Here, we use Tidal, the Haskell-based live coding pattern language. For a complete description of Tidal, see [4] and tidal.lurk.org. The language is designed to facilitate ease of composition in live performances, and allows the user to quickly generate complex manipulations of synthesised sound. To get a flavour of how patterns are coded in Tidal, we describe the components of the pattern from the bottom of Fig. 2:

```
d3 $ palindrome $ rev $ every 4 (iter 12) $ iter 2
$ jux (jux (striate 9)) $ slow 8
$ sound"sundance*30" |+| vowel "a o a"
```

The **d3** at the beginning specifies that sound will be sent to the 3rd of the 9 channels in the Dirt synthesiser. A series of *pattern transformer* operations are applied. Each $ sign applies the operator to everything that follows, as if the whole of the remainder of the pattern were in parentheses. These operators change the speed, order and sub-sampling of the pattern of samples that is specified by the **sound** operator, which is the only mandatory command. The **sound** command has a single argument: a pattern of samples enclosed within quote marks. In this example, the sample **sundance** is repeated 30 times. There are many different ways to specify when the sample will be played within a prespecified time period (default 1 cycle per second). Following from the sound command, a series of synthesiser parameters can be used, separated by the |+| operator. In this example, the vowel command uses a formant filter to mask the sample with a vowel-like sound. The vowels used are *also* specified by a pattern, in this case "a o a". The pattern of samples is controlled by a pattern of vowel shapes, and each pattern is organised in time independently of the other. This combination of transformations can radically change the way these samples can sound to the human ear.

2.2   Grammar and Parser

There are strong arguments for running the evolutionary algorithm externally to the Tidal process. Firstly, it is good design practice to keep the code generation/manipulation functionality independent from the code execution – this is safer because it the former will not interfere with the scheduling of the latter. Secondly, it means we can run the grammar tools as if they were acting as another composer. Finally, it means that we have a separate representation of the grammar that we can manipulate in order to yield attractive patterns.

Constructing the Grammar. There are many ways to represent a language in a grammar. Note that our aim here was not to create a grammar that could parse all *possible* Tidal patterns, but rather to create a functional grammar that could parse the majority of patterns and allow us to implement some mutation functions to act upon that reduced grammar, mainly because we wanted to test the whole system before expending too much time refining the grammar. Accordingly, we created a grammar that was sufficiently representative of patterns that live coding use. This is not an exhaustive grammar, but the majority of the structures in the test corpus patterns could be parsed. It was more important that syntactically illegal patterns were excluded from the evolutionary system.

We used two sources of pattern data to construct the grammar: the Tidal reference pages at tidal.lurk.org; and a set of example patterns available from yaxu.org/tmp/patternlib/. The former was used to ensure that all the documented commands in Tidal were represented in the grammar, and that the syntax from the examples was correctly implemented. The latter was used to test the ability of the grammar to parse the more sophisticated constructs that can be created in Tidal. We call this data set the *test corpus*.

We used Antlr [8] to generate a java library from the grammar we developed. A subset of the parsing rules are shown in Fig. 4 in the Antlr grammar format. This library was then used to build the parse functions needed to generate patterns automatically, and also to build the mutation operators for the evolutionary algorithm.

One of the advantages of working with live coding systems is that they are designed to be 'crash-proof': badly formed patterns do not break the system, they merely generate an error message as feedback to the composer. This is

```
trans_spec
: trans_0arg
| LBRK trans_spec  RBRK (POINT LBRK trans_spec RBRK)*
| slow_pattern
| STUT INTEGER zero2one (zero2one | LBRK MINUS zero2one RBRK)
| SUPERIMP LBRK trans_spec RBRK
| cont_frag
| (DEG_BY|TRUNC) zero2one
| (number)? (BEATR|BEATL)
| int_arg_trans intint
| EVERY INTEGER ((LBRK trans_spec RBRK)| trans_spec)
| FOLDEVERY LSQB INTEGER (COMMA INTEGER)+ RSQB LBRK trans_spec RBRK
| (SOMETIMESBY_ALIASES | SOMETIMESBY zero2one) trans_spec
| WHENMOD INTEGER INTEGER LBRK trans_spec RBRK
| WITHIN LBRK zero2one COMMA zero2one RBRK LBRK (trans_spec|(KNIT cont_frag )) RBRK
| JUX LBRK trans_spec (POINT trans_spec)* RBRK
| ZOOM LBRK zero2one COMMA zero2one RBRK
| STRIATE1 INTEGER LBRK (ONE|INTEGER) DIVID (ONE|INTEGER) RBRK
| SMASH intint LSQB number (COMMA number)* RSQB
| slowspread_pattern ;
```

Fig. 4. Example of rules used to construct the Tidal parsing grammar

analogous to biological systems, where expressed enzymes cannot change the laws of chemistry (c.f. crash the system), but can be sufficiently harmful (c.f. generate an error message) to alert the organism that evolved them to respond appropriately.

2.3 Evolutionary Algorithm

Our goal is to prove the concept of evolving DSL-based pattern music in a live setting. Accordingly, we developed a minimal genetic algorithm, capable of generating mutated patterns from a population of patterns on demand.

Our grammar-based genetic algorithm uses the parser described above to evolve the population. Since the parser can tokenise any legal Tidal pattern, it is possible to use the patterns themselves as the genotype of the evolutionary process, since mutation is also handled by the parser. In this respect, our approach differs from Grammatical Evolution [9], which uses a sequence of integers to generate a pattern from the parse tree. Our approach is more akin to Genetic Programming, but is able to handle different values of terminal nodes due to the use of specific mutation operations for particular node classes.

Default values for the parameters of the evolutionary algorithm are shown in Fig. 5. Genetic algorithms have five stages. The first is an *initialisation* stage, followed by an iteration through stages of *selection, crossover, mutation* and *evaluation*. Finally, a *termination* step is introduced to decide when to stop.

Variable	Value	Notes
Mrate	0.4	Mutation rate
maxpop	20	Maximum number of individuals in the population

Fig. 5. Default values for the parameters in the system

Each of these stages requires special consideration in a live coding setting. For ease of reference, we detail each of these stages below.

Initialisation. Although there are many potential strategies for automatically initialising a population, we felt that it was important for the coder to select patterns to submit to the algorithm from the outset. There are two reasons for this. Firstly, it means that mutations will always have basis in the tastes of the coder. Secondly, it means that we did not have to implement a full pattern generator via the grammar (this will be the subject of future work).

Selection. Our selection strategies work in two ways. Positive selection is based on the fitness of the individual. When the coder uses the 'pull' command to select a pattern from the population, the chance of a pattern being selected is proportional to fitness. When a new pattern is added to the population, one of the patterns with the minimum current fitness is automatically deleted.

Crossover. Crossover is a controversial topic in genetic algorithms, and it was not necessary to implement it in our current algorithm. We don't do crossover automatically yet, but we propose that if we only swap over identical node types, then crossover can never be fatal. That is the advantage of a grammar-based mutation strategy. Coders have the opportunity to cut and paste fragments of mutated patterns should they see fit, but there is no pressure to do so.

Mutation. To test our approach, we implemented three mutation mechanisms which operated on three different regions of a Tidal pattern: insertion of pattern and sample transformers (e.g. the phrase `$ every 4 (iter 12)` for the pattern in Fig. 2); substitution of sample patterns (e.g. `sundance` cold be changed for `jvbass` or `[bd cp jvbass]`; and insertion of synthesiser parameters (e.g. `|+|`
`vowel "a e o i"`). (For more details of this syntax, see tidal.lurk.org). Mutations were positioned in the pattern by identifying the appropriate nodes in the parse tree that the mutations could be applied to. This guarantees that the mutations are always syntactically correct, and allows complicated nesting of sample patterns and transformations to emerge through evolution. Given the lack of crossover, it was important that the mutation operator could insert new branches into the parse tree. This is achieved in two ways: insertion of transformations and synthesiser patterns; and insertion of new groups of pattern events in a single event (for example, mutating the pattern `"sn sn bd sn"` to `"sn [hc ho hc] bd sn"`).

As in standard GAs, 'harmful' mutations are bound to emerge. The 'damage' that they will do in this application is evaluated by the coder - they are sounds which are in some way incompatible with the performance at the current time. Also, some mutations are neutral. For example, if a 'delayfeedback' command is created, it will have no effect unless the 'delay' parameter is set. Since this will do no harm (other than to make the pattern text more obscure), there is no real problem if these things emerge. The coder, observing such mutations, might consider to turn these features 'on' by adding the necessary text, feeding this back in to the evolving pattern population.

It is possible to produce many modifications of the same class, for example |+| vowel "a e i" |+| vowel "a e o u i". The Tidal interpreter will superimpose these transformations on each other, placing them at appropriate positions through the time period. In this example, the first pattern will be spread out at a period of 1/3 of the cycle time and the second pattern will be spread out over 1/5 of the cycle time.

Mutation is a stochastic event. Since our genetic algorithm is interactive, and mutations are requested by the user, it is important that a change *usually* happens. However, it is also important that the mutation is not so great that there is no relationship between parent and offspring. When a coder presses 'pullmut', they are *requesting* a new mutation, so they don't want to get something back that they've seen already – there is much more emphasis on novelty. We experimented with a range of mutation rates, and further work here is needed, but we found that setting the probability of 0.4 per mutable node gave a good balance of stability and innovation. The number of mutable nodes varies depending upon the sample pattern, so the mutation rate will change between different patterns. Every pattern contains at least one **message** node, and at least one **sample** terminal node, so the probability p_p of mutation per pattern is $p_p \geq \sqrt{0.4}$.

Although we have shown how mutation happens when the 'pullmut' button is pressed, it is important to emphasise that the coder continues to edit the evolving patterns throughout the performance, and could be considered to be another sort of mutation or crossover operator. This is one of the strengths of the approach we have developed, as it allows a more flexible interaction between the coder and the algorithm.

Evaluation. Evaluation is the assignment of fitness to patterns. The perceived fitness of a pattern is dependent on the current patterns being interpreted into audio at any one time. In this application, fitness is determined by the buttons that the coder presses to evaluate a pattern and interact with the current population of patterns. For any pattern in an input box that also exists in the genome, when a button is pressed the fitness will be changed by the following values:

- **eval:** +0.5
- **push:** +0.5 (if pattern already exists in the population)
- **pull:** 0
- **pullmut:** +0.49
- **up:** +0.25
- **dn:** −0.25

Termination. The choice of when to end a performance is up to the ensemble of coders. It is possible that the ensemble will stop using the GA towards the end of a performance in order to craft a satisfying ending to the performance by hand.

3   Evalutation

Source code for our evolvable version of Extramuros is available at github.com/anon/Extramuros. Source code for the ParserMutator server, which runs on Tomcat 7, is available at github.com/anon/ParserMutator. For clarity, the concept we are testing is the use of evolved patterns during a live coding performance, which proceeds as follows:

1. the Extramuros system is initialised as normal
2. the ParserMutator web server is initialised
3. the coder(s) specifies the web client url in the browser
4. the coder(s) enters the password.
5. Begin live coding with evolution as in Fig. 3

The following sections detail the evaluations we have carried out on the system.

3.1   Is the System Generating Legal Tidal Patterns?

Yes. The mutations we designed were a subset of all possible instances of the grammar, and could always be parsed by the Tidal interpreter. Note though that not all hand-coded patterns can be parsed as legal by the ParserMutator, as it only recognises a subset of all possible Tidal patterns. For example, patterns using Haskell's applicative functor notation are not currently recognised by our parser, but they can still be submitted to the Tidal interpreter during a performance. We plan to extend our parser in the future so that (almost) all Tidal patterns can be recognised.

3.2   What Is the Relative Level of Complexity of the Generated Patterns?

Figure 6 (top) shows the parse tree for the following evolved pattern:

```
d3 $ palindrome $ every 5 (degrade) $ palindrome $ striate 8 $
sound "feelfx speechless seawolf [tech tink [birds3 gabbalouder tabla2
incoming]] f" |+| delaytime "0.39 0.64 0.33" |+| speed "10.7 0.62"
|+| end sinewave1
```

Figure 6 (bottom) shows the parse tree for the following hand-coded pattern:

```
d1 $ every 2 (slow 2) $ superimpose (iter 4) $ slow 2 $
every 8 (striate 9) $ sound "future*3 bd*2 wobble [cp bd bd]"
|+| speed "[2, 3, 4]" |+| cutoff (slow 16 sine1)
|+| resonance (slow 12 triwave1)
```

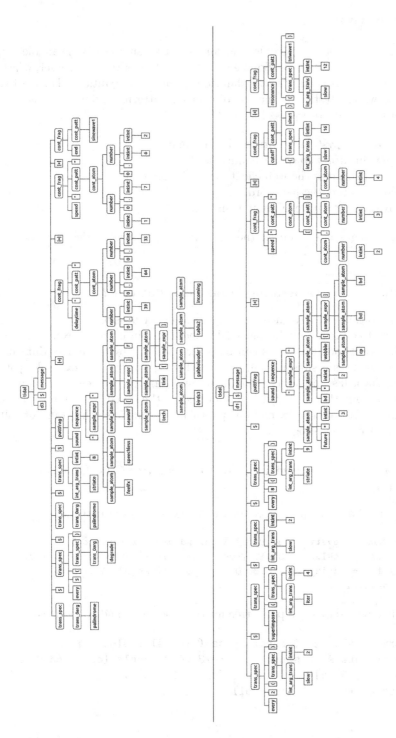

Fig. 6. Parse tree of an evolved pattern (top) and a hand-coded pattern from the Yaxu corpus (bottom).

These diagrams of the parsing process show that the structures are intuitively similar. Both are representative of the level of complexity that live-coded patterns tend to reach, particularly if they are the only patterns that the synthesiser is processing at the time (i.e. no other channels are being processed simultaneously). Patterns are usually in three sections: *transition specifications, pattern specifications,* and *continuous modifiers.* Our mutation mechanisms create parse nodes in each of these areas, and so yield similar parse trees to hand-coded patterns. (A more rigorous comparison will be the subject of future work).

3.3   How Does the Generated Audio Compare with Tidal?

Several videos of the system are available on the (anonymous) YouTube channel www.youtube.com/channel/UCu0_2dIhFvfKV-c975snmXQ. In each of these, the 'pullmut' function was the main source of innovation in the patterns, with the occasional hand-coding or editing of evolved patterns by the user.

These performances compare favourably with hand-coded Extramuros-based performances such as www.youtube.com/watch?v=zLR02FQDqOM, especially if it is considered that the users of the evolved system have little or no previous experience of live coding. We currently have no means of quantitatively comparing live coding performances, which will be the subject of future work.

3.4   How Does the Fitness of the Population Change?

A plot of the change fitness values of a population of evolved patterns is shown in Fig. 7. Early on in the run, there are few patterns in the population because the coder has not added many at this point. Average fitness increases until the

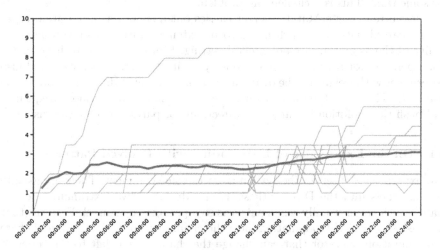

Fig. 7. Change in fitness for evolved patterns. x-axis is the performance time (hh:mm:ss), y-axis is fitness. The fitness of the individual patterns is shown in blue and mean fitness is in red (Color figure online).

5th min. and then drops off as more new patterns with low initial fitness values are added to the population. An individual pattern rapidly increases in fitness as it is used to create new patterns via mutation. The population reaches the size limit of 20 at around the 15th min. From this point when a new individual is added, unfit genomes are removed. Fitness rises from this point onward.

4   Conclusion

We have demonstrated a new system for evolving live coded music in collaboration with human coders. The novelty of the system is that it exploits the networked sharing of live code to allow interaction with an evolutionary music generation module. Having linked a live coding framework to an evolutionary algorithm, we have a convenient system for experimentation with a range of evolutionary algorithms. Several avenues for further research naturally present themselves.

At the moment, a population of patterns exists on each browser client. Whist this is convenient, and allows different coders to maintain their own population of patterns, it might be better to maintain a larger population on the server side, or to allow "horizontal gene transfer" between populations of patterns.

The method of assigning fitness values based on the interaction of the coder with the mutated patterns deserves further attention. We plan to evaluate a range of different fitness scoring strategies within this approach. A key problem that needs to be addressed is the relatively low fitness of newly-generated patterns compared with patterns that exist in the population. One could envisage 'inheriting' a proportion of fitness from the parent pattern to counteract the relatively high fitness of patterns that have been established in the population for some time. This is a challenging problem.

The current ParserMutator server carries out a small set of mutation types on the parsed patterns, which needs to be extended further to investigate the utility of the system in a more general setting. Firstly, we need to have mutation operators for a wider range of node types in the parse tree. We could also experiment with weighting the distribution of mutations at different nodes in the grammar. This would allow us to make certain changes occur more frequently and push the evolution in particular directions for particular applications.

4.1   On the Experience of Live Coding with Evolved Patterns

Our experience as users of the system is worth some comment. There are usually nine channels into the Dirt synthesiser, and during a live Extramuros performance, each coder usually sends patterns to a pre-agreed channel. Patterns that are pushed to the GA population have the channel encoded in them, and there is no mutation operator that can change the channel. It is left to the coder to change the channel specification by manual editing. It might be better to formalise the way that the channels are organised. At the moment they are stored on the genome, but this may not be the best way to arrange things.

Working with the GA as a single user, one tends to spread out patterns across multiple windows - it is as if we are using the GA as one or more extra performers, and allowing it to populate several text boxes with evolved patterns (see video, linked in Sect. 3.3). Although this is different to the convention used by coders in Extramuros, the two approaches can be made to be compatible by prior consent.

Audiences at live coding events are very open-minded, and used to 'glitches' appearing as the performance continues. As long as unsympathetic patterns are removed, than that is accepted as part of the experience. These are the ideal conditions for experiments with music evolution.

The only technical problem we encountered with the evolved patterns was due to differences in sample length in the library of samples available to the Dirt synthesiser. Some samples were short such as the bass drum "bd", whereas other samples were longer, for example "bev", which samples a few seconds of singing. The problem arises when the pattern repeats the long sample in such a way that overloads the memory available to Dirt, resulting in buffer overruns. This reveals one issue where the live coder has an advantage over the genetic process - familiarity with the sample library prevents the coder making these errors. With experience, it is possible for the coder to spot where an evolved pattern is beginning to flood the Dirt buffer, and edit the pattern before the buffer is full. Related to this, our system is useful for novice Tidal coders unused to the nuances and range of functions available in the Tidal language, as the system generates examples of how the various features of Tidal can be used.

4.2   On the Need for a Quantitative Evaluation

Our goal with this contribution was to evolve Tidal patterns for use in a live coding performance. We have offered a qualitative evaluation of this work, but recognise that a quantitative evaluation is desirable, particularly as we implement the improvements discussed above. There are two routes to obtaining a qualitative evaluation, which we sketch out here. The first approach is to make a comparison of the statistics of the patterns. For example we could compare the graphs of the evolved patterns with hand-coded patterns quantitatively using graph statistics, as we did subjectively in Sect. 3.2. The second approach is more challenging, which is to somehow evaluate the artistic qualities of the generated piece. One could apply audio statistics to this problem, but this does not truly capture artistic merit unless it is done by reference to high-quality pieces of music in a similar style. As Fernandez and Vico [1] point out, this is a challenging problem for all automated music generation. It is worth remembering that a quantitative evaluation also offers a route to unsupervised generation of live coding patterns, since it can be used as a fitness function.

4.3   Closing Remarks

By making the connection between the musical process and evolution, we have necessarily made a more direct connection between the parser, the patterns themselves, and the audio that is produced. The audio is in effect part of the phenotype of the pattern. This is very Pattee-like [10] – the actual specification is spread across all representations - function, pattern and parser. It is also worth stating the analogy with biological systems. Biology exists in time. It is not static, and must respond to a dynamic, changing environment. One of the attractions of working in musical systems is that time must be considered implicitly, and the dynamics of the performance must be accommodated.

Acknowledgements. This work was funded by the EU FP7 project EvoEvo, grant number 610427.

References

1. Rodriguez, J.D.F., Vico, F.J.: AI methods in algorithmic composition: a comprehensive survey. J. Artif. Intell. Res. **48**, 513–582 (2013)
2. Unehara, M., Onisawa, T.: Composition of music using human evaluation. In: The 10th IEEE International Conference on Fuzzy Systems, 2001, vol. 3, pp. 1203–1206. IEEE (2001)
3. MacCallum, R.M., Mauch, M., Burt, A., Leroi, A.M.: Evolution of music by public choice. Proc. Nat. Acad. Sci. **109**(30), 12081–12086 (2012)
4. McLean, A.: Making programming languages to dance to: live coding with Tidal. In: Proceedings of the 2nd ACM SIGPLAN International Workshop on Functional Art, Music, Modeling and Design, pp. 63–70. ACM (2014)
5. Ogborn, D., Tsabary, E., Jarvis, I., Cardenas, A., McLean, A.: Extramuros: making music in a browser-based, language-neutral collaborative live coding environment. In: McLean, A., Magnusson, T., Ng, K., Knotts, S., Armitage, J. (eds.) Proceedings of the First International Conference on Live Coding, University of Leeds, ICSRiM, p. 300 (2015)
6. Aaron, S., Blackwell, A.F.: From Sonic Pi to overtone: creative musical experiences with domain-specific and functional languages. In: Proceedings of the first ACM SIGPLAN Workshop on Functional Art, Music, Modeling and Design, pp. 35–46. ACM (2013)
7. Wang, G., Fiebrink, R., Cook, P.R.: Combining analysis and synthesis in the ChucK programming language. In: Proceedings of the International Computer Music Conference, pp. 35–42 (2007)
8. Parr, T.: The Definitive ANTLR 4 Reference. Pragmatic Bookshelf, Dallas (2013)
9. O'Neill, M., Ryan, C.: Grammatical Evolution: Evolutionary Automatic Programming in an Arbitrary Language, vol. 4. Springer, New York (2012)
10. Pattee, H.H.: Cell psychology: an evolutionary approach to the symbol-matter problem. In: LAWS, LANGUAGE and LIFE. Springer, pp. 165–179 (2012)

Towards Adaptive Evolutionary Architecture

Sebastian Hølt Bak, Nina Rask, and Sebastian Risi$^{(\boxtimes)}$

IT University of Copenhagen, Copenhagen, Denmark
{sbak,ninc,sebr}@itu.dk

Abstract. This paper presents first results from an interdisciplinary project, in which the fields of architecture, philosophy and artificial life are combined to explore possible futures of architecture. Through an interactive evolutionary installation, called *EvoCurtain*, we investigate aspects of how living in the future could occur, if built spaces could evolve and adapt alongside inhabitants. As such, present study explores the interdisciplinary possibilities in utilizing computational power to co-create with users and generate designs based on human input. We argue that this could lead to the development of designs tailored to the individual preferences of inhabitants, changing the roles of architects and designers entirely. *Architecture-as-it-could-be* is a philosophical approach conducted through artistic methods to anticipate the technological futures of human-centered development within architect.

Keywords: Evolutionary computation · Interactive evolution · Architecture · Artificial life

1 Introduction

With inspiration in artificial life *'as-it-could-be'* [1,2] present paper seek a synthesis that extends exploration and research into the domain of architecture *'as-it-could-be'*. The ability to create unexpected situations and the opportunity to be rooted in biological, life-like occurrences is a key ingredient in our artistic and phenomenological approach to investigating the relationship between humans and future built space. The paper concerns itself with artificial evolution as a parametric design tool for architects and its use is investigated by way of a performative method conducted through an installation in a real life living lab; the results are analyzed by way of phenomenology.

Our approach revolves around the installation *EvoCurtain* (Fig. 1), which explores the use of *interactive evolutionary computation* (IEC; [3]) as user-driven processes for configuring architectural elements. The installation is meant to be a showcase of how IEC could aid in designing and evolving architecture along with occupants from parameters defined by architects and designers. The goal of this research is to examine co-evolutionary possibilities between humans and technology in the context of built space. We investigate the ability to share autonomy with something technological and study what impact evolving spatial qualities has on the human experience of inhabiting a particular space.

C. Johnson et al. (Eds.): EvoMUSART 2016, LNCS 9596, pp. 47–62, 2016.
DOI: 10.1007/978-3-319-31008-4_4

Fig. 1. EvoCurtain Installation. The visual patterns projected onto the curtains is able to adapt to the preferences of the inhabitant through interactive evolution.

Computational technologies are already being used as tools for creating designs and calculating structures within the field of architecture [4–6]. New materials, construction methods and design directions have arisen from the impact of computers and technology. However, with our installation we are more interested in exploring the possibility for computers to work together with or assist users in creating designs for their individual preference. For the architectural practice, the challenge with the future use of computer-aided design arises when it is required to achieve a more dynamic and site-specific outcome, as the computational design-tools currently operate with a high degree of automation.

When architects are expected to design parameters rather than final designs they are placed in a completely new role. This new role demands the creation of works driven by dynamic potential instead of the traditional design knowledge. In relation, Galanter [7] mentions Whitelaws notion of "metacreation" that puts the artist in charge of designing the initial process, which is later left to fulfill its own inherent prospective. Combined with the interactive aspect of IEC, we suggest that the key to attaining future living conditions with great individual appeal lies in the ability to design a potential, which enables inhabitants to relate to the space they inhabit in a new way.

A new branch of architecture that is specifically invested in interactive, evolving and site-specific buildings is *adaptive architecture* [8,9]. The field is concerned with buildings that are designed to adapt to input from its surroundings; examples range from art installations to zero energy houses [10], where input can come from the inhabitants, environmental data or other key areas. Common to all projects is the notion of buildings and spaces as being flexible, interactive, dynamic or responsive, which is where the field of *artificial life* (alife) becomes an apt supplement.

Our artistic exploration of desired phenomena shares kinship with approaches within alife, as it is conducted by (artificial) stagings of various phenomena – in our case, the phenomenon of humans sharing a living space with an evolutionary algorithm. From an interdisciplinary perspective, we argue that the particular fields involved in our investigation can benefit from a greater mutual involvement in future research.

1.1 Research Design and Goal

The investigations in this paper are part of a larger interdisciplinary research project concerning adaptive architecture, wherein an approach known as *the growth plan* was employed. The growth plan is a method for innovative development of technology, which consists of three cyclical phases of exploration [11,12]. We appropriate this approach in three phases: (1) An initial exploration phase, (2) a lab prototyping phase, and (3) a real life evaluation phase.

In the final phase, which this paper focuses on, the adaptive capabilities and temporal aspects of alife become an integral part of research and implementation in a real life setting. As seen in Fig. 2, we sought to encompass the complexity of real life by combining alife with disciplines pertaining to built space and human experience. The challenge of investigating and enabling co-existence and mutuality between humans and adaptive technologies is approached as socio-technical. The possible applications and experiences of our installation are evaluated with experiential performance methods [13]. The aim of combining the various complementary disciplines in an explorative process is for meaning to emerge in the context-specific interaction between EvoCurtain and its participants. To examine co-evolutionary possibilities between humans and technology in the context

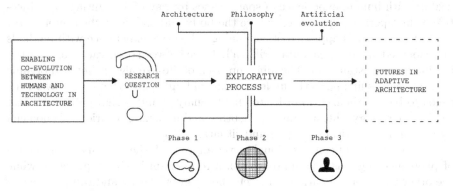

Fig. 2. Project Overview. Knowledge and experiments are generated by means of a process that progresses through the three cyclical phases of Growth Plan [11]. In order to attain an explorative study of the possible future of architecture, the research was not approached from an initial hypothesis, but rather evolved from an initial research question. In order to analyse the growth through these phases, a framework and vocabulary that enables an interpretation of human experiences is attained through phenomenology, which in turn is situated in a context of architecture and artificial life in order to contribute to the interdisciplinary field of adaptive architecture.

of built space, it is imperative to employ a tool that can aid in investigating the inhabitants' experience of engaging in a reciprocal relationship with the habitat itself.

It is furthermore important to note that the goal of EvoCurtain is not the generation of new designs, but rather to be able to longitudinally evaluate the impact and experience of the installation in a real life setting. This enables us to investigate the transformational and behavioral effects on humans in a holistic setting. And although the goal is not to develop new designs, such an approach creates opportunities to evaluate the potential of this interdisciplinary combination with regards to future approaches.

2   Background

This section reviews relevant methodologies that the approach presented in this paper builds upon.

2.1   Interdisciplinary Scope: Philosophy, Architecture and ALife

In order to understand and analyse the experiences with EvoCurtain, we looked to phenomenology - the interpretive study of human experience. Simply put, the phenomena and issues pertaining to future adaptive architecture are not readily available to our experience and investigation. Human experience can be amplified by way of technical installations, and as we seek to understand experiences mediated by technology not yet existing, this approach aids in altering experience by way of challenging life in various environments. In this context, Merleau-Ponty's phenomenology of embodiment [14] serves great purpose as its dealing with human experience of space comes by way of accounting for subjective or first-person experiences, where the body and user is at the center.

Merleau-Ponty's approach of combining subjective experience with methods more focused on empirical data strikes a balance between observation and participation in the installation. The alife approach of life-like algorithms and organic environments is fitting, as the understanding of built spaces that evolve over time needs to be externalized in order to enable analysis and design potentials. Just as phenomenology, alife attempts to capture and understand various phenomena by way of hypothetical examples and situations [15].

Having the aid of computational power aligns with the imaginative approach of phenomenology, as one approach takes root in subjective experience, while the other employs simulations - thereby being interrelating and complementary approaches. And when entering areas of future studies, where various data and experiences are not readily available for research, the conceptual approach of architecture as-it-could-be becomes a useful tool - not just in a philosophical sense as a technological extension of subjectivity, but as an explorative tool for designing potential interactions with built space.

In the presented installation, alife gives our living room presence and agency, enabling us to engage in the complexity of real life situations that could occur when living alongside an evolving curtain.

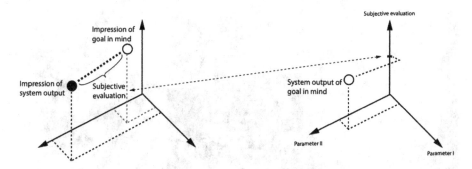

Fig. 3. Interactive Evolution. This figure shows a simplified version of Takagi's model [3]. The goal in mind of the user is situated together with the impression of the system output. The difference between impression and goal in mind form the subjective evaluation of the user, which then becomes the third parameter that the program adjusts after, forming its candidates for next evaluation.

2.2 Interactive Evolution

Participants in the presented study can change curtain designs through IEC [3], which uses human feedback to evaluate content. In IEC, human users make aesthetic decisions by rating individual candidates, thereby deciding which individuals breed and which ones die instead of relying on fitness functions designed by developers (Fig. 3). IEC has been applied to several domains such as music and sounds [16–19], images [20], and visuals for video games [21–24]. While IEC has shown promise in a variety of domains, it has not yet been applied in the conventional forefront of architectural design.

3   EvoCurtain: The Evolutionary Window Curtain

The installation EvoCurtain explores the use of interactive evolution as user-driven processes for configuring architectural elements. The installation is meant to be a showcase of how IEC can aid in designing and evolving architecture along with occupants from parameters defined by architects and designers. To explore architecture-as-it-could-be through an IEC approach, the artifact needs to integrate with the apartment and be perceived as a natural extension of existing features in the home - features that would then adapt to the inhabitants preferences over time.

Additionally, the presented project aims to merge important capabilities of humans, houses and computational technologies: The human ability to relate and sense the world through their body and perceive it subjectively, the ability of the house to provide not only protection and shelter, but also mental space for defining quality of life, and finally the ability of technology to connect, gather, compute and transform data, connecting the human as well as the house to their surrounding environment.

Fig. 4. EvoCurtain. Similarly to a membrane or curtain, EvoCurtain allows users to filter the view to the image of the outside world.

This notion resulted in a decision of letting the computer subject the inhabitant to images from the surrounding city in which the apartment was situated. These images are projected onto white, curtain-like pieces of hanging fabric and are covered by a pattern of white dots (Fig. 4). The pattern is generated by IEC and functions as a membrane or curtain, filtering the view to the image of the outside world.

3.1   Set-Up and Spatial Layout

The experiment is set up among all the existing furniture in the living room of the apartment (Fig. 5). Two fabric walls are installed, which consist of floor-to-ceiling

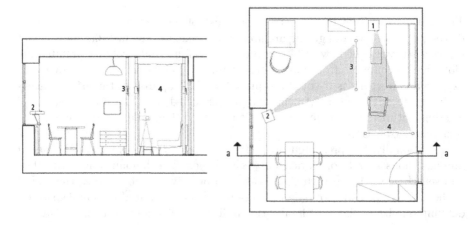

Fig. 5. Section a-a and plan of living room showing the projections of EvoCurtain and placement of wall screens.

poles with white curtains. Two computers connected to two projectors display the image on the fabric walls, turning them into a resemblance of windows.

3.2 Genetic Representation

It is important that the curtain allows the inhabitants to see more or less of the background image depending on their preferences. Five parameters are able to evolve through IEC: the minimum size of the ellipses, the maximum size of the ellipses, the speed with which they go from min to max, the amount of ellipses in total and finally their distribution over the area of the curtain. The color and transparency of the dots remains static. The ellipses are able to increase and decrease their size with varying speed and with the influence of a randomization function they appear to flutter, similarly to a curtain that waves in the wind.

These parameters form a way for the curtain to intervene by acting like a changing membrane between the inhabitant and the image. The variable ranges were tweaked and experimented with in order to create a design that would both be able to contain enough variation for the inhabitant to evaluate, but that would also resemble a natural looking element in the apartment, like a curtain.

3.3 Interactive Evolutionary Setup

The algorithm that generates EvoCurtain is an example of IEC designed to have a significant impact on its surroundings. When the algorithm initializes, it generates a population of five members. One member is shown at a time, allowing the inhabitant to rate it between zero and nine. The algorithm generates five curtains per day, and each curtain should have at least two to three hours to be experienced, before moving on to the next. When the present member is evaluated, the next member in line is shown. The members of the population thus get assigned fitness depending on the rating given by the inhabitant.

Based on the evaluation of the five members, the algorithm assigns fitness scores that decide how much chance each one has of becoming one of the parents for the next generation. The genetic properties of the two chosen parents are then combined and mutated to create the next generation of five members, to be evaluated in the following session.

4   Experiments

The experiment was conducted over the course of ten days, where an interventional installation was set up in an apartment home to one of the authors. The experiment included ourselves (two of the authors) as well as four additional test participants through an experiential method. The method used in the experiment was conducted by way of three different sessions, exploring the space in both a personal and social way.

Following the *triangulation method* [13], We performed three different sessions, each represented a different angle on the space that we were exploring

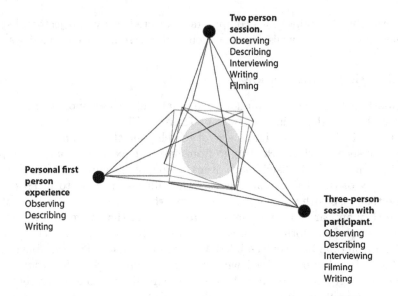

Two person session.
Observing
Describing
Interviewing
Writing
Filming

Personal first person experience
Observing
Describing
Writing

Three-person session with participant.
Observing
Describing
Interviewing
Filming
Writing

Fig. 6. Observational structure. The experiment is observed through three different angles.

(Fig. 6). The first session, which continued throughout the whole experiment, consisted of ourselves experiencing the room through living with it for ten days, noting down our own first-person experiences. The second was 20 min interviews where we each took turn to question each other and get questioned in focused two-person sessions. The third session was a structured performative three-person session, where we invited test-participants from the first phase and assigned each a role of instructor, observer and experiencer - assuming the same triangular set of observation positions as in our second phase of experimentation. An important supplement to the method was that the experiencer was equipped with a camera in order to record their 1st person-perspective. This was done with test-participants as well as when we questioned each other.

The video from all perspectives would afterwards be re-watched and interpreted by us. We had furthermore added an extra criterion in our questioning: When continuously asking the participants about their relation to the room, the relation was to be accompanied by an evaluation, which was then plotted into the evolving algorithm. This was to create a recurrence of relation that brought not only the co-evolving engagement to the forefront, but also the question of who was in control.

Outside of the staged methods various scheduled activities such as lunch with guests and other 'everyday' activities were added in order to give depth to the first-person experience for the one of us was living in the space.

4.1   Genetic Experimental Parameters

Each of the five genes describing the evolving curtain had a 25 % chance to mutate every time a new generation is created. The minimum size of the ellipses was set between 0 and 70 pixels. The maximum size was set between the minimum and 70 pixels. The number of ellipses could vary between 100 and 1000 and the position of each ellipse was generated by a perlin noise function. The speed of the ellipses was scaled between zero and one, where zero is a complete standstill and a setting of one means pulsating between the max and min size once per second. The parameter ranges were tweaked and experimented with in order to create a design that would both be able to contain sufficient variation for the inhabitant to evaluate, but that would also resemble a natural looking element in the apartment, like a curtain.

5   Results

The algorithm aided in providing a sincere experience of living with something that had intelligence and agency - a computer program manifested as a visual prototype, which was capable of reacting on its own and interpreting inputs. The fact that the algorithm were convincing became evident when one of our participants referred to it as something she could have discussions with and afterwards commentated on its changing behavior by saying *"Oh so you think you know what I want, huh?"* Participants furthermore wondered how this relation would be if living together with another human and the algorithm. Their thoughts on this was that the algorithm would act as a kind of conciliator between the human inhabitants in that the participant imagined that they would have to all adapt to each other to co-exist.

Test participants provided a necessary perspective on our own experiences, as the need of assuming several observational positions gave us additional insight into how a living situation with multiple agents with agency could occur and thereby qualified our own role in the experiment.

Figure 7 shows an example of the progression of the interactive evolution guided by a test participant. Evolution takes place when genes of members from current generations mix and form the future generations of patterns. The change thus unfolds without inhabitants noticing and only when the representation emerges for them to experience and evaluate, do they see the result of the evolutionary process. The fact that anything is changing is sometimes not even visible at first; only after a few generations do the test participants discover the change. The mixing of genes, combined with mutation, evolve into more and more similar patterns based on the evaluation of the human. It must be noted here that the patterns were not static in nature and thus still images do not do it favor. A video of EvoCurtain in action can be found here: https://goo.gl/d8ITBz.

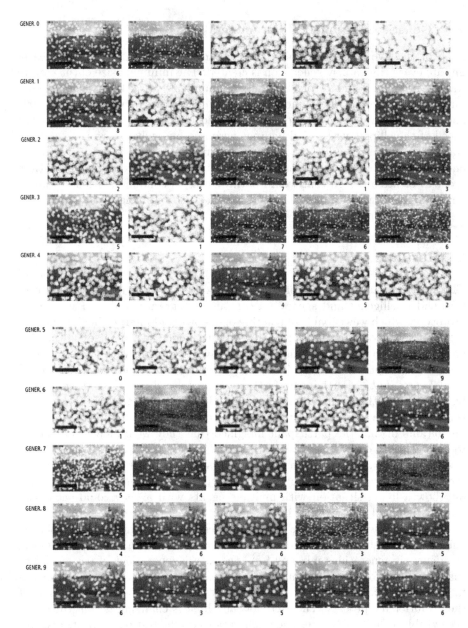

Fig. 7. An example of ten generations with five members each that were rated by a test participant.

5.1 Disagreement and Reconciliation

Following the first stages of the experiment, the algorithm reached a point where it appeared 'stubborn' – it kept showing similar patterns during the span of five generations regardless of the chosen patterns. This was observed as the inhabitants and the algorithm 'disagreeing' on the spatial qualities of the room.

As time went by, test participants started to have less focus on the patterns as a purely visual feature and started experiencing it as a metaphor of something living. This marked a certain shift in the way the room was engaged and the knowledge that became embodied in that relation helped re-open the space to interpretation. The use of IEC over a longer period of time meant that we were allowed to study evolving relationships between humans and built space, which changed and achieved a certain rhythm as it became more integrated in everyday activities. It was no longer a question of controlling EvoCurtain, rather, the attitude of participants tended towards allowing oneself to co-evolve with it. The advantage of having the prototype develop in a longitudinal study thus became clear, as the experiences of who was in control drastically changed. This had not occurred in earlier experiments, where prototypes were only engaged for a maximum of four hours.

5.2 Control and Preconceptions

Despite initial concerns about the experience of relinquishing control of one's intimate space to something non-human, it ceased to present itself as a major challenge. Our interpretation of what the participants had experienced in earlier non-alife experiments was that their lack of control made them uncomfortable; they regarded the prototypes as potentially non-cooperative.

We argue that the issue for the participants in the second phase was that they had to come to terms with the technology as an 'other', to use Ihde's terms [25], and realise that how they moved and interacted was a 'dialogue' - a relationship that needed time to develop. It was realized that an important thing to consider is the preconceptions humans have about technology: We are used to controllable and carefully designed interaction. Having to live with technology that in unfinished in the beginning, but develops over time is a new relation that humans will most likely need to adapt to in the future. A simplified but current example of such relation is the iPhone's Touch ID, which improves its ability to recognize the user's fingerprint over time. The fact that the identification does not work perfectly at first but after a few days functions seamlessly, might represent an anticipation of such a phenomena. In the future, technology might develop alongside humans instead of coming out of the box with a finished design, as the advantages of letting it develop and adjust to individuals is much greater than that of uniform designs.

5.3 Time

The temporal aspect was an underlying theme for the entirety of our project. The perceived changes in the relation to EvoCurtain meant that the participants' experience of inhabiting an adaptive space felt more akin to being *with* the room, rather than merely in it. During the longer period of testing with EvoCurtain, it was observed how the room slowly ceased to be regarded as an object and was instead experienced as something dynamic and living. We thereby believe it is possible to argue for the inherent quality in developing future living environments, if approached as an entity that adapts its design qualities through an evolving relationship with the inhabitant based on individual preferences, as opposed to a building with customizable smart technology.

5.4   Interdisciplinary Perspective

It is no trivial matter to evaluate how successful the application of IEC and alife is in an interdisciplinary project like EvoCurtain. A measurement of IEC's utility and performance is challenging to achieve in an objective manner, as the results are based on user interaction and subjective experiences. However, the fact that computer simulations helped stage some of the complex environments pertaining to adaptive architecture, not only allowed for an investigation of human experience in realistic settings, but speaks to an initial success from an interdisciplinary perspective.

As such, using artificial evolution to explore the possibilities of adaptive architecture appears to have many complementary qualities. Just as a gradual adaption over several generations can be called evolution, adaptive architecture can come to be called (interactive) evolutionary architecture. Prototypes with adaptive capabilities seem to be a necessary component when exploring the design of living technology and spaces to be inhabited in a holistic manner. Using evolutionary algorithms in explorations of how built space can be understood and designed in a meaningful way for the individual has significant implications for architecture (spatial design), arts (staging of experience) as well as philosophy (analysing the experience of the space). The initial estimate is thus that the privileged position of each field can aid in yielding tools, which meet the requirements for designing built spaces that has dynamic and adaptive interpretations of the surroundings, where both human and surrounding environments are included in the loop.

6   Discussion

Langton suggests that *"traditional biology studies life-as-we-know-it, (...) Alife seeks to explore the possibilities of life-as-it-could-be"* [26]. In this paper, we argue that alife can be used to explore architecture-as-it-could-be.

Alife explores phenomena using artificial means to expand our imagination of what is possible and IEC can create an outcome that is based on human input. Architects are concerned with the quality of living spaces, and evolutionary algorithms are concerned with fitness values. From the perspective of the inhabitant it is one and the same: The built environment that is best fit to be currently inhabited. What is important to note here, is the interim nature of the definition – the same space might be perfect in one situation and unsuitable in another. Thus neither evolution nor humans gravitate towards finalized solutions, but rather the most fit proposal in a given situation. To architects, this is a new way of looking at computation; through the eyes of nature. Evolution takes time and "problem solving" in nature is a slow process. Development of buildings in the future could be thought of as an adaptive process guided by the interplay between inhabitant and building. This means that there are no predefined goals, as the aim of adaptive architecture is to adapt to the inhabitants and the surroundings as the current preferences, needs, and conditions develop and change.

Jaskiewicz [27] considers the idea of stretching the sketching phase of a building project to continue into the actual built structure, providing the same adaptive features as the computer or architectural cardboard models: *"The architectural models developed as representations of future habitats could be developed in such a way, that their inherent logics are employed to drive the adaptive processes in actual architectural spaces"*. Through his thorough analysis of the architectural practice and the functional programming of buildings, Jaskiewicz concludes that the ability to adapt to its environments is an important success criteria for architecture: *"Architectural spaces that can be frequently transformed to match changes in the conditions of their internal and external, artificial and natural environment are bound to be more "successful" than those spaces that do not adapt to these changes"*. He further goes on to say that this criteria is something upon which buildings should be evaluated, as it relates closely to the holistic performance of buildings and therefore the success as living spaces.

What we found in our studies was that humans in turn would have to adapt themselves to the system in order to enable the interaction that constantly makes both parties better suited for each other. Adaptive architecture is as such an adaption in (1) the built structure, (2) the mind of the inhabitant and (3) the role of the architect. Considering Whitelaw's [7] notion of metacreation, the shift in the future role of architects and designers can be suggested to have much in common with the shift happening in the role of the artist: *"Metacreation refers to the role of the artist shifting from the creation of artifacts to the creation of processes that in turn create artifacts"*. In *vernacular architecture* [28] buildings change as extra rooms and new kitchens are created from existing mass, morphing out where space is available and building upwards where it is not. This notion should be kept in mind when imagining buildings that evolve based on the feedback of their users, in which only the overall process is defined by an architect. We suggest that computer technology can enable an extension of imagination, using means that aid sense, eyesight and logic, while being evaluated by humans to push exploration and development in a direction most fit for human life to unfold. Jaskiewicz [27] mentions evolution as a safe strategy for evolving adaptive buildings. He suggests that in order to solve the problem of adaptive architecture evolving into something dangerous we must look at the complex systems of biology. Here, the two main qualities are development and evolution, and over time adaptive architecture will become increasingly more secure and dependable. He also argues that a way to avoid initial challenges would be to make use of virtual scenarios for testing, before implementations in real life. This notion suggests that when inhabitants conduct the fitness evaluation it is possible to assume that no building evolves in an unwanted or harmful way. Following our investigations, we noted that the question of who is in control was relatively non-pertinent. Rather, it became imperative that inhabitants were able to perceive themselves as participants and creators rather than users, and to expect a mutual development to unfold as opposed to using a finished product. In order to enable inhabitants in actively engaging with future built spaces, one must understand how the relation to an adaptive building can be made meaningful for the individual in an everyday setting.

Through our investigations on adaptive architecture, the contribution that alife offers in an interdisciplinary setting becomes apparent. In trying to understand human experience in the context of possible future technologies, the use of alife helped create environments that introduced a sense of embodiment and engagement. It furthermore elucidated how simulations can lead to new experiments when investigating complex human experiences. EvoCurtain has obvious limitations regarding scale, time and objectivity when it comes to investigating human navigation and experience of adaptive spatial qualities. We thereby argue that in future investigations of this kind, philosophy and architecture alike can benefit from alife's ability to construct phenomena and data that can aid in the development of new designs based on more advanced understandings of human-technology relations.

Adaptive architecture thus poses a new challenge in developing interdisciplinary methodologies for designing and evaluating parameters in a holistic manner that appeal to our embodied way of being. We argue that an interdisciplinary and explorative approach akin to the small steps taken in this project is key to anticipate the futures of adaptive evolutionary architecture: architecture-as-it-could-be.

7 Conclusion

Our work presented and evaluated the implementation of IEC in the context of architecture and philosophy. This interdisciplinary approach is represented in the EvoCurtain installation and has proven a useful approach in attempting to unfold the complexities inherent to adaptive architecture. The way EvoCurtain was implemented can in itself present a novel approach to how future adaptive environments are to be investigated, where the qualities of the building adapt to the activity and behavior of its inhabitants, be it lighting, spatial qualities, thermal conditions or other conditions pertaining to the spaces we inhabit.

Due to the nature of the installation it was difficult to measure the direct impact of IEC, we have however seen indications of the relevancy of employing alife and IEC to attain aspects of the complex experiences pertaining to the inhabitation of an adaptive environment. As it became apparent how alife amplified the nature of the relationship to the built environment, it not only enabled more refined philosophical studies of how participants experienced the surroundings, it offered a way of anticipating the possibilities in future relations between inhabitants and built space.

References

1. Bedau, M.A., McCaskill, J.S., Packard, N.H., Rasmussen, S., Adami, C., Green, D.G., Ikegami, T., Kaneko, K., Ray, T.S.: Open problems in artificial life. Artif. Life 6(4), 363–376 (2000)
2. Langton, C.G.: Artificial Life: An Overview. MIT Press, Cambridge (1997)
3. Takagi, H.: Interactive evolutionary computation: Fusion of the capabilities of EC optimization and human evaluation. Proc. IEEE 89(9), 1275–1296 (2001)
4. Kalay, Y.E.: Architecture's New Media: Principles, Theories, and Methods of Computer-Aided Design. MIT Press, Cambridge (2004)
5. Wiberg, M.: Interactive Textures for Architecture and Landscaping: Digital Elements and Technologies: Digital Elements and Technologies. IGI Global, Hershey (2010)
6. Schumacher, P.: Parametricism: a new global style for architecture and urban design. Archit. Des. 79(4), 14–23 (2009)
7. Galanter, P.: The problem with evolutionary art is. In: Di Chio, C., et al. (eds.) EvoApplications 2010, Part II. LNCS, vol. 6025, pp. 321–330. Springer, Heidelberg (2010)
8. Schnädelbach, H.: Adaptive architecture-a conceptual framework. In: Proceedings of Media City (2010)
9. Schnädelbach, H., Glover, K., Irune, A.A.: Exobuilding: breathing life into architecture. In: Proceedings of the 6th Nordic Conference on Human-Computer Interaction: Extending Boundaries, pp. 442–451. ACM (2010)
10. Farahi Bouzanjani, B., Leach, N., Huang, A., Fox, M.: Alloplastic architecture: The design of an interactive tensegrity structure. In: Adaptive Architecture, Proceedings of the 33rd Annual Conference of the Association for Computer Aided Design in Architecture (ACADIA). ACADIA, Cambridge, Ontario, Canada, pp. 129–136, October 2013
11. Ross, P., Tomico, O.: The growth plan: An approach for considering social implications in ambient intelligent system design. In: Proceedings of the AISB 2009 Convention, pp. 6–9. Citeseer (2009)
12. Magielse, R., Ross, P., Rao, S., Ozcelebi, T., Jaramillo, P., Amft, O.: An interdisciplinary approach to designing adaptive lighting environments. In: 2011 7th International Conference on Intelligent Environments (IE), pp. 17–24. IEEE(2011)
13. Petersen, K.Y., Søndergaard, K.: Staging multi-modal explorative research using formalised techniques of performing arts. In: KSIG 2011: SkinDeep - Experiential Knowledge and Multi Sensory Communication (2011)
14. Merleau-Ponty, M., Smith, C.: Phenomenology of Perception. Motilal Banarsidass Publisherm, Delhi (1996)
15. Froese, T., Gallagher, S.: Phenomenology and artificial life: toward a technological supplementation of phenomenological methodology. Husserl Stud. 26(2), 83–106 (2010)
16. Hoover, A.K., Szerlip, P.A., Stanley, K.O.: Functional scaffolding for composing additional musical voices. Comput. Music J. 38(4), 80–99 (2014)
17. Dahlstedt, P.: Creating and exploring huge parameter spaces: interactive evolution as a tool for sound generation. In: Proceedings of the 2001 International Computer Music Conference, pp. 235–242 (2001)
18. Johnson, C.G.: Exploring sound-space with interactive genetic algorithms. Leonardo 36(1), 51–54 (2003)

19. Jónsson, B.T., Hoover, A.K., Risi, S.: Interactively evolving compositional sound synthesis networks. In: Proceedings of the 2015 Annual Conference on Genetic and Evolutionary Computation, GECCO 2015, pp. 321–328. ACM, New York (2015)

20. Secretan, J., Beato, N., D'Ambrosio, D.B., Rodriguez, A., Campbell, A., Folsom-Kovarik, J.T., Stanley, K.O.: Picbreeder: a case study in collaborative evolutionary exploration of design space. Evol. Comput. **19**(3), 373–403 (2011)

21. Risi, S., Lehman, J., D'Ambrosio, D.B., Hall, R., Stanley, K.O.: Combining search-based procedural content generation and social gaming in the petalz video game. In: Proceedings of the Artificial Intelligence and Interactive Digital Entertainment Conference (AIIDE 2012), Menlo Park, CA. AAAI Press (2012)

22. Hastings, E.J., Guha, R.K., Stanley, K.O.: Evolving content in the galactic arms race video game. In: Proceedings of the IEEE Symposium on Computational Intelligence and Games (CIG 2009), pp. 241–248. IEEE (2009)

23. Hoover, A.K., Cachia, W., Liapis, A., Yannakakis, G.N.: Audioinspace: exploring the creative fusion of generative audio, visuals and gameplay. In: Johnson, C., Carballal, A., Correia, J. (eds.) EvoMUSART 2015. LNCS, vol. 9027, pp. 101–112. Springer, Heidelberg (2015)

24. Risi, S., Lehman, J., D'Ambrosio, D., Hall, R., Stanley, K.: Petalz: Search-based procedural content generation for the casual gamer. IEEE Trans. Comput. Intell. AI Games **PP**(99), 1 (2015)

25. Ihde, D.: Bodies in Technology, vol. 5. University of Minnesota Press, Minneapolis (2002)

26. Keeley, B.L.: Artificial life for philosophers. Philos. Psychol. **11**(2), 251–260 (1998)

27. Jaskiewicz, T.: Towards a methodology for complex adaptive interactive architecture. TU Delft, Delft University of Technology (2013)

28. Oliver, P.: Encyclopedia of Vernacular Architecture of the World. Cambridge University Press, Cambridge (1997)

Plecto: A Low-Level Interactive Genetic Algorithm for the Evolution of Audio

Steffan Ianigro[1]([✉]) and Oliver Bown[2]

[1] Architecture, Design and Planning, University of Sydney,
Darlington, NSW 2008, Australia
steffanianigro@gmail.com
[2] Art and Design, University of New South Wales,
Paddington, NSW 2021, Australia
o.bown@unsw.edu.au

Abstract. The creative potential of Genetic Algorithms (GAs) has been explored by many musicians who attempt to harness the unbound possibilities for creative search evident in nature. Within this paper, we investigate the possibility of using Continuous Time Recurrent Neural Networks (CTRNNs) as an evolvable low-level audio synthesis structure, affording users access to a vast creative search space of audio possibilities. Specifically, we explore some initial GA designs through the development of *Plecto* (see www.plecto.io), a creative tool that evolves CTRNNs for the discovery of audio. We have found that the evolution of CTRNNs offers some interesting prospects for audio exploration and present some design considerations for the implementation of such a system.

Keywords: Neural network · Genetic Algorithm · Evolution · Interaction design

1 Introduction

The prospect of evolution as a creative tool for the generation of music has been explored by many artists through the authoring of Genetic Algorithms (GAs) such as *Living Melodies* by Palle Dahlstedt and Mats G. Nordahl [1]. These GAs are highly abstract biological models and attempt to harness the unbound possibilities for creative search evident in nature. However, when appropriating this principle into the world of digital audio, creators of these systems are confronted with many design challenges, such as how to represent audio within the system. For example, McCormack [2] emphasises how vast and unsearchable the space of possibilities is when evolving a 200,000 pixel image. Even though McCormack refers to brute-force random search, the same notion is true for the space of possibilities when evolving an audio waveform by manipulating its individual samples. Although this approach could potentially afford almost limitless creative possibilities, it would be an impractical solution as the vastness of the GA's search space would hinder creative use.

C. Johnson et al. (Eds.): EvoMUSART 2016, LNCS 9596, pp. 63–78, 2016.
DOI: 10.1007/978-3-319-31008-4_5

Many programmers have explored alternatives by creating audio representations within GAs, evolving formalised systems for audio creation that they think will yield interesting creative results [3], such as a synthesis structure. This means that the GA's functionality can shift from low-level to high-level, which we define as:

- **Low-level** - low level of abstraction between genotype and phenotype, e.g. the phenotype is an audio waveform and the individual samples of the waveform are treated as the genotype.
- **High-level** - high level of abstraction between genotype and phenotype, e.g. the genotype encodes a synthesis structure that is then used to produce the audio waveform phenotype.

As a result, the system's output is constrained within the possibilities afforded by the audio engine's higher-level components as individuals produced by the system will exhibit strong traits of the underlying formalised structures that created them [3]. The scope of the system's search space is reduced, meaning the output is more manageable to explore, thus creatively useful. It is, however, hard to ignore the creative potential of adopting lower-level alternatives which strip away high-level audio engine structures and explode the audio search space. Yet this potential is of little use without a viable low-level audio engine, raising the question: What is a viable evolvable low-level audio engine that offers users a vast and explorable creative search space? If a user just wants to sift through parameter permutations of a software synthesiser for an appealing setting, this question may be of little interest, however we feel exploration of lower-level alternatives could prove valuable in seeking compositional possibilities outside a user's immediate resource.

Within this paper we look into how Continuous Time Recurrent Neural Networks (CTRNNs) can be used as an evolvable low-level audio engine as well as what affordances such a system has compared to rigid high-level alternatives. Bown [4] identifies the benefits of a human-centred approach to designing systems that produce 'creative' output. This approach moves beyond algorithmic functionality and considers subjective aspects of user-interaction with a system to achieve a design that successfully aids creativity. We adopt this stance in the creation of some initial designs for *Plecto* (see www.plecto.io), a GA which evolves a CTRNN audio engine.

2   Background

2.1   Genetic Algorithms

Genetic Algorithms are based on Darwinian theory, with evolutionary change a result of the fittest of each generation surviving and passing on the traits that made them fit [5]. This process starts with the random generation of an initial pool of candidates. This population is evolved through various iterations of a genetic process during which each individual's fitness is evaluated in relation to a selective pressure. The fittest individuals are bred through a crossover process during which genetic traits from both partners are passed onto their offspring. There is also a chance of mutation as to allow genetic drift or exploration outside the current gene pool. The new individuals replace a number of the least fit existing individuals and the process is repeated [6]. Asexual reproduction is also possible, with mutation being the only force for change within the population.

GAs provide a powerful method for searching a problem space, optimising candidates until the best solution is found. The selective pressure of a GA usually exists as an automated fitness function, evaluating individuals according to a criterion that is encoded into the system. Some GAs also incorporate human evaluation known as Interactive Genetic Algorithms (IGAs) and are of most interest to the authors. The predominant paradigm for IGAs sees the user take the role of a 'pigeon breeder', acting as a selective pressure in an artificial environment [7]. This is an appealing prospect as it is difficult to define explicit fitness functions for audio phenotypes that can identify subjective musicality [8]. Incorporation of human interaction shifts the GA into a new domain, as it "... is no longer a tool for finding optimal solutions, but instead becomes a vehicle for creative exploration [9, p. 31]." *Plecto* adopts an IGA structure inspired by *Picbreeder* [10], allowing users to selectively evolve preferred individuals. An open-ended evolutionary system is achieved, absent of explicit fitness objectives, in an attempt to reproduce the unbound innovation of natural evolution [11].

2.2   Creative Potential of Lower-Level Audio Engines

In [12, 13], Yee-King describes two GAs used to evolve audio. Both papers describe systems that attempt to match an incoming audio stream by evolving an audio engine in real time. The selective pressure of these GAs is a fitness function that compares the GA's output to its target sound. The system described in the first paper could be considered high-level, as the user chooses any existing VSTi (Virtual Studio Technology instrument) as an audio engine for the GA to evolve. Therefore, the system's output is constrained within the possibilities of the high-level formalised structures that make up the selected VSTi. For example, Yee-King describes the system as being able to replicate synthesiser sounds but not sounds produced by acoustic instruments [12], a product of the spectral complexity of acoustic instruments falling outside the selected VSTi's capabilities. The second paper by Yee-King proposes a similar system although adopts a lower-level audio engine. The genotypes for this GA exist as data describing the number of oscillators and parameters of a synthesis algorithm. A mechanism to grow the genome is also discussed, achieving a malleable audio engine that can change in structure to match a more complex target sound [13]. Contrasting with the system discussed in [12], the GA is not limited to rigid high-level structures, but can adapt to achieve greater variety and complexity in the audio it produces due to its lower-level design.

Darwin Tunes [14] is another example of a GA used to evolve audio, and allows users to influence the evolution of various audio loops. The system's audio engine has computational layers consisting of 'pre-composed' sound agents that can be configured and arranged in various ways. When a member of the population is previewed, the audio engine is reconfigured according to the individual's genome data which specifies the arrangement and configuration of these sound agents. Once the audio is rendered, the user can listen to the audio loop and steer the genetic process depending on their response by rating selected individuals of the population, increasing or decreasing their fitness and chance of survival. *Darwin Tunes* can be considered high-level as the audio engine has predetermined musical structures such as a set tempo of 125BPM, a time signature of 4/4, and the use of the twelve note scale standard in Western musical tradition. These musical limitations constrain the system's output, however are purposeful so that users are not deterred by too much variation in the population. Instead, *Darwin Tunes* achieves variety in the population through a lower-level approach to instrumentation. Rather than adopting preset sounds such as piano or guitar, additive synthesis is used to create timbral content from scratch. A more unfamiliar audio space results, helping remove emotional responses to certain instruments and expectations of their behaviour that may cause bias during population evaluation [14]. This is again a positive implication of a lower-level audio engine structure, helping remove cultural influence as a variable in this artificial ecosystem.

3   A Low-Level Audio Engine

Within the literature, there is good reason to believe that as high-level structures are removed from GA audio engines, their potential for creative search is increased. However, even though audio engines such as the additive synthesis algorithm discussed in [13] are capable of extensive diversity, we question if there are lower-level options that offer greater potential for creative search. In [15], Magnus presents an alternative, describing a system which evolves audio waveforms and strips away many of the meta-relationships between genotype and phenotype. Magnus defines each gene as a segment of a waveform between two zero crossings. When an individual of the population is bred with another, they swap genetic traits (waveform segments), producing a new combination. There is also a small chance of mutation of the waveform segment. The purpose of this system is not to search for completely new sonic material but instead "...to produce a series of electroacoustic pieces in which the ecological process of evolution is transparent in the form of the piece" [15, p. 1]. This is an interesting prospect, however not ideal for our purpose as with no generator for new timbral content, the creative search space of the GA would be limited to arrangements of existing audio material. Therefore, it is evident that there is a need for some sort of audio engine to produce divergence in the search space. But what evolvable audio engine could be used that functions on a lower level than those described in [12–14]?

Within the literature there are many examples where neural networks have been used in musical contexts. In [16], Biles et al. propose the use of a Feed Forward Neural Network as a fitness function, training it to identify 'good' or 'bad' musical output. In [17], Mozer describes *CONCERT*, a system which implements a neural network to create novel musical compositions. *CONCERT* is trained by a set of existing compositions and produces output based on the initial material. In [18], Bown and Lexer discuss a system which utilises a CTRNN as a generator of discrete time events at a control rate in response to a performer's input. These events are used to trigger musical output, making the system a direct participant in a musical performance. Bown and Lexer also outline the possibility of using CTRNNs as an evolvable audio synthesis structure. The notion of using a Neural Network to produce audio is also discussed by Ohya [19] who explores the possibility of training a Recurrent Neural Network on an existing piece of audio and then altering its parameters to synthesise variants of the original sound. Eldridge [20] also explores the use of Continuous Time Neural Models for audio synthesis, identifying that these chaotic systems are "...capable of producing evolving timbres reminiscent of natural sounds which are hard to achieve by conventional synthesis techniques"[20, p. 1]. This prospect inspired *Plecto*.

CTRNNs can exhibit complex temporal behaviours and are a simple nonlinear continuous dynamical model, capable of approximating trajectories of any smooth dynamical system [21]. Due to this, they are well suited to produce audio output as various configurations result in smooth oscillations that resemble audio waveforms. CTRNNs are very different to the aforementioned audio engine examples of evolved VSTis and other synthesis architectures as their subsymbolic nature alleviates the restrictive design factors of formalised high-level audio structures. They are an interconnected network of computer modelled neurons, typically of a type called the *leaky integrator*. For this research, we adopt a fully connected CTRNN, meaning that the neurons in the hidden layer are all connected (including a self connection) and the input layer has a full set of connections to the hidden layer [18]. We also calculate the output or activation of each node using a *tanh* transfer function, providing node outputs between −1 and 1 for use as samples in an audio wavetable. The behaviour of a neuron is defined by three parameters - Gain, Bias and Time Constant - and each connection between neurons has a Weight parameter that governs the strength of neuron interactions within the CTRNN. Due to the complex relationship between network parameters and behaviour, GAs are often used to calibrate CTRNNs for a specific task, such as discussed in [22].

Plecto adopts a CTRNN as its audio engine, providing an evolvable low-level audio generator capable of extensive diversity. In this initial stage of development, *Plecto* manipulates the Gain, Bias, Time Constant and Weight Parameters of a CTRNN with one input neuron and twenty hidden neurons. The amount of neurons in the network is maintained to achieve a constant as we focus on experimenting with other parameters. Asexual reproduction is also adopted, as more drastic genetic changes achieved through a crossover process can shift the CTRNN into a fitness trough. Once a CTRNN configuration is found that produces oscillations, an audio buffer can be filled with the CTRNN output for playback once DC offset of the waveform is corrected. This audio buffer can also be converted into any audio encoding format such as WAV, allowing users to import the generated audio into any Digital Audio Workstation (DAW). Upon implementation of this CTRNN audio engine, we are now confronted with the difficulties of exploring an IGA with an extremely large search space. In the remainder of this paper we develop a framework embedded in interaction design to investigate how discovery and search of this vast creative search space can be possible.

4   Growing the Population

In [23], Machado et al. identify a problem that results when human evaluation replaces machine evaluation of a GA's population. The cumbersome process of evaluating each individual creates fatigue, preventing prolific exploration of a large problem space as a single person would only discover a small segment of the system's sonic possibilities. Therefore, we are faced with a problem of effectively discovering the population's potential while maintaining user influence over the evolutionary process. The notion of collaborative GAs discussed in [9] provides avenues for overcoming this problem, and refers to systems that are explorable by multiple users. Time and space considerations of physical collaborative paradigms become irrelevant; the cumbersome task of evolving a vast population is shared and is no longer the responsibility of a lone user. *Darwin Tunes* [14] provides an example, demonstrating evolution by public choice. It implements evolution in a complex social environment, where fitness is defined by the aesthetic tastes of the system's community of users. *Picbreeder* [10] is another example of a system which taps into the creativity of large online communities to overcome the fatigue of single-user evaluation, while still maintaining subjective evaluation of the population. The benefits of collaborative evolution is demonstrated by the variety and size of the populations of both *Darwin Tunes* and *Picbreeder*; therefore we feel the incorporation of a means for multi-user interaction is a valuable design consideration when developing IGAs that evolve low-level audio engines.

Plecto adopts this collaborative functionality, existing as a web-based environment open to multiple users. *Plecto*'s client to server communications are based on the structure of the systems described in [9,24]. Information is exchanged via HTTP protocol with a central server. All the audio processing is completed on the client side, meaning only small chunks of information need to be passed to and from the server that describe the client side audio engine configuration. While this information exists as small score files in [24], and as a string of integers in [9], in *Plecto* this information exists as a Javascript object containing CTRNN configurations which are rendered as audio on the client side once requested from the server. As new configurations are found by users, they are saved to this remotely stored object that is accessible by any *Plecto* user. We have established the technical specifics of how multi-user interaction is implemented, but what of the interface providing multiple users' control over the evolutionary process?

Both *Darwin Tunes* [14] and *Picbreeder* [10] provide different methods of allowing multi-user influence over the system's evolutionary process. *Darwin*

Tunes provides a less direct method in which users rate individuals according to aesthetic preferences, increasing or decreasing their chance of survival within an automated genetic process. *Picbreeder*, on the other hand, exhibits more direct user interaction, allowing users to hand breed each individual privately before contributing a new image to the population. *Plecto* adopts a similar method of user interaction to that in *Picbreeder*, featuring an interface inspired by *Picbreeder*'s 'Tree of life Browser'. A family tree of individuals is depicted that links parents to children, allowing users to privately select individuals of interest and mutate them towards a desired aesthetic. Once a pleasing individual is found, a user can save it to the public population which others can access. This approach is adopted as it helps users understand the evolutionary process and how traits are passed on to new individuals as opposed to the more automated approach of *Darwin Tunes*. Users can also navigate back through the generations and create new family lines if the genetic path taken does not prove fruitful.

Upon investigation of this interface for evolving individuals, it is evident that the process of mutation can be tedious, meaning only a few individuals are contributed to the main population per user session. If thousands of users are engaging with *Plecto*, rapid progress will be made to discover the population's potential, but this may not be a realistic use case. We strive to build a system that allows users to navigate a vast population of evolved individuals. To generate this vast population, we speculate that a more autonomous approach may be more effective, such as an interactive ecosystem model. *Eden* [25] is an example, consisting of autonomous agents which utilise sound to assist in survival. McCormack describes the artwork as 'reactive', meaning a "...viewers' participation and activity within the physical space around the work have important consequences over the development of the virtual environment" [25, p. 3]. *Plecto* could feature a similar dynamic, with population divergence influenced by more abstract user input, a prospect we aim to explore in future versions.

5   Searching the Population

5.1   Meaningful Representation of the Population

As *Plecto* currently stands, we have implemented a collaborative IGA structure that taps into online communities to increase the rate of population exploration. We have explored some possibilities of visually arranging the population when privately evolving individuals but this interface will become cluttered if *Plecto*'s whole public population is displayed. Therefore, a different interface design is needed that affords coherent depiction of *Plecto*'s whole population so that users can discover the rewards of multi-user interaction. In [26,27], the authors discuss the benefits of using visualisations as a means to understand the population of a GA. Many of these visualisations are based on comparative fitness measures of individuals of a population, providing insight into the fitness landscape and optimisation effectiveness of the GA. The subjective nature of evaluating individuals of an IGA makes similar visualisations more difficult as the fitness measures of individuals will vary from user to user depending on their creative requirements, removing a constant point of comparison required to visually plot the population. This point is where many IGAs lose effectiveness, as once interesting populations are created, they become hard to navigate and fully utilise as their interfaces are of little help in understanding characteristics of the population, thus making the process of search one of chance. This is a problem accentuated in *Plecto* as we are now dealing with an extremely vast population of a low-level audio engine. As data visualisations of standard GAs prove effective, we ask how similar population visualisations can be bridged to the subjective world of IGAs?

5.2   Creating GA Landscapes

An interesting example of data visualisation can be found in [28], in which a game-like system used to explore music collections called *deepTune* is discussed. This system offers a 3D virtual landscape consisting of tracks of a music collection, arranged according to audio analysis information and statistical data of each track. Similar songs appear close together in the landscape, with greater distance separating songs that have little resemblance (see Fig. 1). An environment full of visual landmarks, or 'islands of music' which guide the user through the musical search space is achieved, providing "...a fun way to explore music collections in virtual landscapes in a game-like manner" [28, p. 1]. Implementing a similar visualisation as *Plecto*'s interface for population navigation could aid in alleviating the problems of current IGA interfaces.

To achieve a similar visualisation to that seen in *deepTune* [28], some form of automated individual comparison is necessary so that the individuals of the population can be meaningfully arranged. In [29], Schwarz proposes a system called *CataRT*, a compositional tool that allows a user to navigate a corpus of sound segments derived from an initial piece of audio. As seen in Fig. 2, each segment is arranged in a scatterplot according to its audio attributes. The X-axis is defined by the Spectral Centroid of each segment, the Y-axis is defined by the Periodicity of each segment, and the colour of each point is defined by the Note of

Fig. 1. Example of *deepTune*'s interface [28]

Fig. 2. Scatterplot of audio segments from *CataRT* [29].

the corresponding segment. This is an effective layout as the structures created depict information about each segment and its relation to others. Audio segments with similar features are closer together in the scatterplot, and segments that differ greatly are separated by a greater distance.

In *Plecto*, we have adopted a similar population visualisation to that seen in *CataRT* [29], providing a more communicative searchable interface. As each individual is produced by *Plecto*'s GA, its phenotype is discretely rendered and various audio analysis processes are conducted. We have used this audio analysis data to arrange *Plecto*'s population into a landscape defined by these audio features, any of which can be plotted on the interface's X or Y axis. For now, as seen is Fig. 3 (left), *Plecto*'s current implementation uses Spectral Centroid data to plot individuals on the X-axis and Spectral Flatness data to plot individuals on the Y-axis. However, we aim to make these audio features definable by a user in future iterations of *Plecto*.

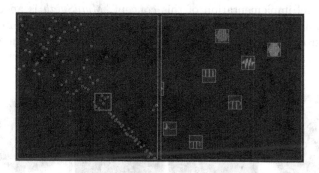

Fig. 3. User interface of [30].

Similarly to *CataRT* [29], the relative positions of individuals within *Plecto*'s current interface provide visual pointers that aid in searching the population. For example, if a user is searching for bright, noisy samples, they can focus their search around the bottom right hand corner of the scatterplot that displays individuals known to exhibit such traits. This scatterplot arrangement also creates clusters such as the funnel shape evident in Fig. 3 (left). These clusters provide visual structures which can be used to maintain user orientation within the population landscape, allowing users to come back to an area if it proves interesting or to avoid revisiting exhausted areas. However, even though presentation of the population as a whole is more effective and meaningful, we feel representations of the individuals themselves could be improved.

5.3   Descriptive Icons

Picbreeder [10] offers an IGA user interface consisting of image icons that are the individuals of *Picbreeder*'s population. When a user navigates the population, it is easy to instantly get an accurate understanding of the characteristics of each individual, as the act of previewing is undertaken in the visual domain (see Fig. 4). *Biomorphs*, discussed in [31], also provide an example of visual individuals, existing as small graphic images drawn by an algorithm which interprets a numeric genotype. In the paradigm of evolving audio artefacts, communicating phenotype characteristics to a user is more problematic as the phenotype leaves the visual domain. This means that an individual cannot be evaluated at a glance, but instead the user has to engage in the time consuming task of listening to the individual's phenotype to gain an understanding of its characteristics. This issue is evident in *Darwin Tunes* [32] in which individuals are represented by circular icons. The only means of gaining an understanding of an individual's characteristics is to listen to its audio phenotype, a cumbersome task compared to the visual approach evident in *Picbreeder*.

In [34], the importance of visual symbols used to represent the growing complexity of computational functions is identified as they represent the capabilities of a system in an efficient way. The user interface of *MutaSynth* discussed in [35] provides a novel implementation of this concept. *MutaSynth* is an IGA which evolves either an exisiting audio engine such as a software synthesiser or an audio engine built from scratch. *MutaSynth*'s genotype maps to the parameters of the selected audio engine, allowing a user to interactively evolve new parameter

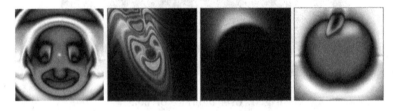

Fig. 4. Visual representations of evolvable individuals in *Picbreeder* [33].

Fig. 5. Visual representations of evolvable individuals showing genome similarities in *MutaSynth* [35].

configurations that may produce more desirable sonic results. Each individual of *MutaSynth*'s population displays a unique symbol derived from its genotype structure. Dahlstedt [35] identifies that these icons provide a visual mechanism for users to identify genome similarities as well as providing visual pointers which can aid a user's memory during navigation of the population (see Fig. 5). Implementing similar visual communication in *Plecto* could aid in visually communicating the characteristics of each individual, increasing the effectiveness of *Plecto*'s scatterplot interface.

We adopt visual icons in *Plecto* inspired by the use of symbols in *MutaSynth* [35]. However, as the visual symbols of *MutaSynth* are derived from individual genomes, nothing is revealed of the phenotype characteristics except for a rough measure of closeness with other individuals. In *Plecto* we explore the possibility of using audio waveform representations as an alternative solution. They occupy a prominent space in digital audio workstations (DAWs) [36] and are also used outside of specialised audio tools, such as appearing in *Sound Cloud* [37] for track navigation. They communicate a generalisation of frequency by the density of the waveform, the volume by its amplitude and even some timbral aspects by its shape (for instance, jagged waveforms indicate spectral complexity). Each waveform also has unique visual attributes (see Fig. 6), potentially achieving a similar dynamic to that discussed in [35], in which icons provide visual stimuli to aid the user's memory during population navigation.

Upon implementation of these waveform icons, it is evident that some insight is provided about phenotype characteristics but this information is very general and only useful on a basic level, such as identifying drastic sonic changes between individuals of the population. In future versions of *Plecto* we will further explore other options such as the more abstract visual icons used in the *Sound Hunters*

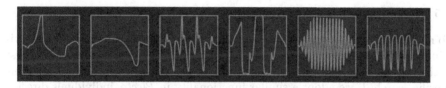

Fig. 6. Icons depicting individuals of *Plecto*'s population [30].

Fig. 7. Icons used in the *SoundHunters* software [38].

[38] software (see Fig. 7). Yu and Slotine [39] identify that spectrograms of different instruments can be used to differentiate them in the visual domain, providing another feasible option. However, regardless of the selected icon type, a problem arises as only so many visual descriptors can be displayed on a single screen and remain perceivable. Therefore to display a large population on *Plecto*'s scatterplot interface, a mechanism for zooming in on areas of the population is required.

5.4 A Navigational Tool for *Plecto*'s Scatterplot Interface

We have identified a method for displaying *Plecto*'s population on a scatterplot as well as possibilities for using symbols to convey information about each individual. To maintain interface coherence as *Plecto*'s population continues to grow, we now propose a navigational tool affording both a global and zoomed in view of the scatterplot interface (see Fig. 3). This navigational tool is inspired by navigational maps evident in many strategy computer games in which an interactive global view of the game world is displayed, allowing a user to shift a more detailed game view to places of interest. In *Plecto*, a small map of the whole population is provided (global scatterplot), with individuals scaled to fit within the map area. Within the complete map exists a smaller square which represents the user's current zoomed in view, showing the icons of each individual. This square can be moved around the map, and as it moves, the user's icon view shifts accordingly. It is evident that *Plecto*'s implementation of this game-like navigation provides a means for users to zoom in on interesting landscape formations and perceive the icons of each individual they wish to explore. However, as *Plecto*'s population grows to the potentially millions of individuals of which its low-level audio engine is capable of producing, the global scatterplot visualisation may become too dense to remain effective. Therefore, we turn to software such as *Google Maps* [40] for a solution.

5.5 Refining Population Search

To navigate *Google Maps* [40], users can filter areas they wish to view by identifying a desired landmark associated with what it is they are looking for. The rest of the map is discarded from view as the user has deemed it irrelevant for their specific search. If we adopt a similar functionality in *Plecto*, individuals directly relevant to a user's creative needs can be filtered from irrelevant candidates of the population and displayed within the scatterplot interface, reducing visual clutter. But what population filtering mechanisms can we use? Some systems adopt categorisation, user ranking and tagging systems to achieve this. *Picbreeder* [10] is an example that allow users to focus population search according to a filter, such as 'top rated', 'most branched' and 'best new images'. However, we question whether there is a more effective filtering solution.

Many types of fitness functions exist in GAs for the creation of audio, from algorithms that simply look for an encoded criterion to algorithms that incorporate machine learning when evaluating individuals of a population [41], such as the trained neural network used to identify fit individuals discussed in [42], or the use of Novelty Search, which rewards divergence in the population [11]. But what if instead of using these fitness functions to optimise the population in the pursuit of a single superior solution, we use them as a tool for arranging the population into meaningful groups? A fitness function could be used to quickly sift through thousands of individuals for a specific criterion identified by a user. In the same way users can filter features on a map by narrowing down the search to restaurants or pubs, users of *Plecto* will be able to search for a specific audio criterion of their choice. In the next iteration of *Plecto*, we aim to adopt this method of using fitness functions as a filtering mechanism. Population subgroups will be assembled by an automated fitness function that identifies user-defined traits. These subgroups will be displayed in the scatterplot interface described above, alleviating visual clutter by creating a lens through which to view the population, aiding in navigating the vast creative search space of *Plecto*'s low-level audio engine.

6   Conclusion

This paper investigates the possibility that greater creative potential lies in GAs that adopt low-level audio engines over rigid high-level alternatives. However, it is unclear what effective low-level audio engine structures exist, and how users can utilise their vast creative search spaces. We have proposed CTRNNs as a possible low-level audio engine and explored some initial system designs through the development of *Plecto* (see www.plecto.io). A collaborative web-based interface is explored that aids in discovering the vast potential output of a low-level audio engine. To better navigate the fruits of multi-user interaction, we propose a method for arranging the population into a visual landscape that provides memorable geography and greater insight into the population's characteristics. The prospect of using visual icons as a filtering mechanism is also proposed, removing the need to engage in the time consuming audio preview process for every individual. So that users can grasp both the fine details of these visual icons and the global view of *Plecto*'s scatterplot interface, we offer a game-like navigational tool that allows users to zoom in and explore interesting sections of the population landscape. The possibility of using fitness functions as another population filtering mechanism is also suggested.

This initial investigation shows promise for the use of CTRNNs as a low-level audio engine. However, as *Plecto* is still in its preliminary stages, further exploration is needed regarding its GA structure and interface design before it is usable as a compositional tool. A problem we aim to address is the lack of diversity in the population, seemingly as a result of the current population being located in a local optimum surrounded by deep valleys of unusable configurations. In *Plecto*'s next iteration we hope to address this by further investigating mutation rates and the possibility of adopting a crossover genetic process. We will also look into achieving a more malleable level of user interaction with the system by providing small CTRNN building blocks which users can assemble to build larger structures in the pursuit of complexity. One possibility for achieving this could be to implement an unsupervised GA that evolves small CTRNN configurations by looking for signs of spectral complexity. Once a large library is built of these CTRNNs, a user can interactively evolve configurations of them, forming a hierarchical system that may afford more interesting audio segments or even full compositions which users can build and utilise within their creative practices. Small CTRNNs could modulate other CTRNNs, similar to frequency modulation algorithms in which oscillators modulate other oscillators. Another option is to fully connect smaller networks together, however the results may prove too temperamental for creative use as the neuron structure of each smaller CTRNN will dramatically change when connected with others. We will also further explore possibilities for conveying *Plecto*'s population as an explorable landscape defined by population characteristics, as well as adopting methods for easier integration into users' existing creative workflows, such as the ability to drag audio samples into existing audio production software.

References

1. Dahlstedt, P., Nordahl, M.G.: Living melodies: coevolution of sonic communication. Leonardo **34**(3), 243–248 (2001)
2. McCormack, J.: Facing the future: evolutionary possibilities for human-machine creativity. In: Romero, J., Machado, P. (eds.) The Art of Artificial Evolution, pp. 417–451. Springer, New York (2008)
3. McCormack, J.: Open problems in evolutionary music and art. In: Rothlauf, F., et al. (eds.) EvoWorkshops 2005. LNCS, vol. 3449, pp. 428–436. Springer, Heidelberg (2005)
4. Bown, O.: Empirically grounding the evaluation of creative systems: incorporating interaction design. In: Proceedings of the Fifth International Conference on Computational Creativity (2014)
5. Husbands, P., Copley, P., Eldridge, A., Mandelis, J.: An introduction to evolutionary computing for musicians. In: Miranda, E.R., Biles, J.A. (eds.) Evolutionary Computer Music, pp. 1–27. Springer, New York (2007)
6. Tzimeas, D., Mangina, E.: Dynamic techniques for genetic algorithm-based music systems. Comput. Music J. **33**(3), 45–60 (2009)
7. Bown, O.: Ecosystem models for real-time generative music: A methodology and framework. In: International Computer Music Conference (Gary Scavone 16 to 21 August 2009), The International Computer Music Association, pp. 537–540, August 2009

8. Tokui, N., Iba, H.: Music composition with interactive evolutionary computation. In: Proceedings of the 3rd International Conference on Generative Art, vol. 17, pp. 215–226 (2000)

9. Woolf, S., Yee-King, M.: Virtual and physical interfaces for collaborative evolution of sound. Contemp. Music Rev. **22**(3), 31–41 (2003)

10. Secretan, J., Beato, N., D'Ambrosio, D.B., Rodriguez, A., Campbell, A., Folsom-Kovarik, J.T., Stanley, K.O.: Picbreeder: a case study in collaborative evolutionary exploration of design space. Evol. Comput. **19**(3), 373–403 (2011)

11. Lehman, J., Stanley, K.O.: Exploiting open-endedness to solve problems through the search for novelty. In: ALIFE, pp. 329–336 (2008)

12. Yee-King, M., Roth, M.: Synthbot: an unsupervised software synthesizer programmer. In: Proceedings of the International Computer Music Conference, Ireland (2008)

13. Yee-King, M.J.: An automated music improviser using a genetic algorithm driven synthesis engine. In: Giacobini, M. (ed.) EvoWorkshops 2007. LNCS, vol. 4448, pp. 567–576. Springer, Heidelberg (2007)

14. MacCallum, R.M., Mauch, M., Burt, A., Leroi, A.M.: Evolution of music by public choice. Proc. Nat. Acad. Sci. **109**(30), 12081–12086 (2012)

15. Magnus, C., Cal IT CRCA: Evolving electroacoustic music: the application of genetic algorithms to time-domain waveforms. In: Proceedings of the 2004 International Computer Music Conference, pp. 173–176. Citeseer (2004)

16. Biles, J., Anderson, P., Loggi, L.: Neural network fitness functions for a musical IGA (1996)

17. Mozer, M.C.: Neural network music composition by prediction: Exploring the benefits of psychoacoustic constraints and multi-scale processing. Connection Sci. **6**(2–3), 247–280 (1994)

18. Bown, O., Lexer, S.: Continuous-time recurrent neural networks for generative and interactive musical performance. In: Rothlauf, F., et al. (eds.) EvoWorkshops 2006. LNCS, vol. 3907, pp. 652–663. Springer, Heidelberg (2006)

19. Ohya, K.: A sound synthesis by recurrent neural network. In: Proceedings of the 1995 International Computer Music Conference, pp. 420–423 (1995)

20. Eldridge, A.: Neural oscillator synthesis: Generating adaptive signals with a continuous-time neural model

21. Beer, R.D.: On the dynamics of small continuous-time recurrent neural networks. Adapt. Behav. **3**(4), 469–509 (1995)

22. Blanco, A., Delgado, M., Pegalajar, M.: A genetic algorithm to obtain the optimal recurrent neural network. Int. J. Approximate Reasoning **23**(1), 67–83 (2000)

23. Machado, P., Martins, T., Amaro, H., Abreu, P.H.: An interface for fitness function design. In: Romero, J., McDermott, J., Correia, J. (eds.) EvoMUSART 2014. LNCS, vol. 8601, pp. 13–25. Springer, Heidelberg (2014)

24. Jordà, S.: Faust music on line: An approach to real-time collective composition on the internet. Leonardo Music J. **9**, 5–12 (1999)

25. McCormack, J.: Evolving sonic ecosystems. Kybernetes **32**(1/2), 184–202 (2003)

26. Routen, T.: Techniques for the visualisation of genetic algorithms. In: Proceedings of the First IEEE Conference on Evolutionary Computation, 1994, IEEE World Congress on Computational Intelligence, pp. 846–851. IEEE (1994)

27. Mach, M.Z., Zetakova, M.: Visualising genetic algorithms: a way through the labyrinth of search space. In: Sincak, P., Vascak, J., Kvasnicak, V., Pospichal, J. (eds.) Intelligent Technologies-Theory and Applications, pp. 279–285. IOS Press, Amsterdam (2002)

28. Schedl, M., Höglinger, C., Knees, P.: Large-scale music exploration in hierarchically organized landscapes using prototypicality information. In: Proceedings of the 1st ACM International Conference on Multimedia Retrieval, p. 8. ACM (2011)

29. Schwarz, D.: The sound space as musical instrument: playing corpus-based concatenative synthesis. New Interfaces for Musical Expression (NIME), pp. 250–253 (2012)

30. Plecto. http://www.plecto.io

31. Nelson, G.L.: Sonomorphs: An application of genetic algorithms to the growth and development of musical organisms. In: Proceedings of the Fourth Biennial Art & Technology Symposium, vol. 155 (1993)

32. Darwin Tunes. http://darwintunes.org

33. Picbreeder. http://www.picbreeder.org

34. Piamonte, D.P.T., Abeysekera, J.D., Ohlsson, K.: Understanding small graphical symbols: a cross-cultural study. Int. J. Ind. Ergon. **27**(6), 399–404 (2001)

35. Dahlstedt, P.: Creating and exploring huge parameter spaces: Interactive evolution as a tool for sound generation. In: Proceedings of the 2001 International Computer Music Conference, pp. 235–242 (2001)

36. Gohlke, K., Hlatky, M., Heise, S., Black, D., Loviscach, J.: Track displays in daw software: beyond waveform views. In: Audio Engineering Society Convention 128, Audio Engineering Society (2010)

37. Sound Cloud. https://soundcloud.com

38. Sound Hunters. http://soundhunters.tv/create

39. Yu, G., Slotine, J.J.: Audio classification from time-frequency texture. arXiv preprint arxiv:0809.4501 (2008)

40. Google Maps. https://www.google.es/maps

41. Bown, O., McCormack, J.: Taming nature: tapping the creative potential of ecosystem models in the arts. Digital Creativity **21**(4), 215–231 (2010)

42. Baluja, S., Pomerleau, D., Jochem, T.: Towards automated artificial evolution for computer-generated images. Connection Sci. **6**(2–3), 325–354 (1994)

Correlation Between Human Aesthetic Judgement and Spatial Complexity Measure

Mohammad Ali Javaheri Javid$^{(\boxtimes)}$, Tim Blackwell, Robert Zimmer, and Mohammad Majidal-Rifaie

Department of Computing, Goldsmiths, University of London,
London SE14 6NW, UK
{m.javaheri,t.blackwell,r.zimmer,m.majid}@gold.ac.uk

Abstract. The quantitative evaluation of order and complexity conforming with human intuitive perception has been at the core of computational notions of aesthetics. Informational theories of aesthetics have taken advantage of entropy in measuring order and complexity of stimuli in relation to their aesthetic value. However entropy fails to discriminate structurally different patterns in a 2D plane. This paper investigates a computational measure of complexity, which is then compared to a results from a previous experimental study on human aesthetic perception in the visual domain. The model is based on the information gain from specifying the spacial distribution of pixels and their uniformity and non-uniformity in an image. The results of the experiments demonstrate the presence of correlations between a spatial complexity measure and the way in which humans are believed to aesthetically appreciate asymmetry. However the experiments failed to provide a significant correlation between the measure and aesthetic judgements of symmetrical images.

Keywords: Human aesthetic judgements · Spatial complexity · Information theory · Symmetry · Complexity

1 Introduction

The advent of computers and subsequent advances in hardware and software, especially the development of tools for interactively creating graphical contents, have turned purely calculating machines into full-fledged artistic tools, as expressive as a brush and canvas. As noted by Michael Noll, one of the early pioneers of computer art, "in the computer, man has created not just an inanimate tool but an intellectual and active creative partner that, when fully exploited, could be used to produce wholly new art forms and possibly new aesthetic experiences" [32, p.89].

Biologically inspired generative tools, especially those utilising evolutionary methods, have been contributed to the creation of various computer generated art with aesthetic qualities. The Biomorphs of Dawkins [12], Mutators of

© Springer International Publishing Switzerland 2016
C. Johnson et al. (Eds.): EvoMUSART 2016, LNCS 9596, pp. 79–91, 2016.
DOI: 10.1007/978-3-319-31008-4_6

Latham [24], and Virtual Creatures of Sims [36] are classic examples of evolutionary art.

An open problem in evolutionary art is to automate aesthetic judgements so that only images of high aesthetic quality are generated [10]. However there are a number of challenges when dealing with evolutionary methods [28]. Therefore, the development of a model of aesthetic judgement model is one of the major challenges in evolutionary art [29] and an essential step for the creation of an autonomous system [27] where both of the generation and evaluation process are integrated.

This paper is organised as follows. In Sect. 2 the relationship between aesthetics and complexity is examined. The notion of complexity from the perspective of Shannon information theory is analysed and its influence on informational theories of aesthetics is discussed. In Sect. 3 we discuss the drawback of entropic approaches for aesthetic evaluation purposes. An in-depth analysis of entropic measure for 2D patterns with examples is provided. In the framework of the objectives of this study, a spatial complexity spectrum is formulated and the potential of information gain as a spatial complexity measure is discussed. In Sect. 4, we provide details of experiments and their results on the correlation of a spatial complexity model with human aesthetic judgements. In Sect. 5, a discussion and a summary of findings is provided.

2   Informational Theories of Aesthetics

Aesthetics has traditionally been a branch of philosophy dealing with the nature of *beauty* in its synthetic forms (i.e. artworks) and its natural forms (e.g. the beauty of a sunset). Computational aesthetics is concerned with the development of computational methods to make human-like aesthetic judgements. The main focus is on developing aesthetic measures as functions which compute the aesthetic value of an object [13]. There a sizeable body of literature on various computational approaches to aesthetics [18,19]. Our review mainly spans models derived from Birkhoff's aesthetic measure and information theory.

Aesthetic judgements have long been hypothesised to be influenced by the degree of order in a stimulus (i.e. symmetry) and the complexity of the stimulus. Birkhoff proposed a mathematical *aesthetic measure* by arguing that the measure of aesthetic quality (M) (Eq. 1) is in direct relation to the degree of *order* (O) and in reverse relation to the *complexity* (C) of an object [9],

$$M = \frac{O}{C}.$$
(1)

The validity of Birkhoff's model, and his definition of order and complexity, has been challenged by empirical studies [39]. Eysenck conducted a series of experiments on Birkhoff's model and suggested that a better expression of aesthetic evaluation function should consider a direct relation to stimulus complexity rather than an inverse relation ($M = O \times C$) [14–16]. Although the validity of Birkhoff's approach in penalising complexity has been challenged by empirical studies, the notion of order and complexity and objective methods to quantify them remains a prominent concern in aesthetic evaluation functions.

Information theory addresses the problem of a reliable communication over an unreliable channel [35]. Entropy is the core of this theory [11]. Let $\mathcal{X}$ be discrete alphabet, X a discrete random variable, $x \in \mathcal{X}$ a particular value of X and $P(x)$ the probability of x. Then the entropy, $H(X)$, is:

$$H(X) = - \sum_{x \in \mathcal{X}} P(x) \log_2 P(x). \tag{2}$$

The quantity H is the average uncertainty in bits, $\log_2(\frac{1}{p})$ associated with X. Entropy can also be interpreted as the average amount of information needed to describe X. The value of entropy is always non-negative and reaches its maximum for the uniform distribution, $\log_2(|\mathcal{X}|)$:

$$0 \leqslant H \leqslant \log_2(|\mathcal{X}|). \tag{3}$$

The lower bound of relation (3) corresponds to a deterministic variable (no uncertainty) and the upper bound corresponds to a maximum uncertainty associated with the random variable. Entropy is regarded as a measure of *order* and *complexity*. A low entropy implies low uncertainty so the message is highly predictable, ordered and less complex. And high entropy implies a high uncertainty, less predictability, highly disordered and complex. These interpretations of entropy provided quantitative means to measure order and complexity of objects in relation to their aesthetic value and consequently contributed to the development *informational aesthetics*, an information-theoretic interpretation of aesthetics.

Moles [30], Bense [6–8] and Arnheim [2–4] were pioneers of the application of entropy to quantify order and complexity in Birkhoff's formula by adapting statistical measure of information in aesthetic objects. Bense argued that aesthetic objects are "vehicles of aesthetical information" where statistical information can quantify the aesthetical information of objects [7]. His informational aesthetics has three basic assumptions. (1) Objects are material carriers of aesthetic state, and such aesthetic states are independent of subjective observers. (2) A particular kind of information is conveyed by the aesthetic state of the object (or process) as *aesthetic information* and (3) objective measure of aesthetic objects is in relation with degree of order and complexity in an object [31].

Herbert Franke put forward a *cybernetic aesthetics* based on *aesthetic perception*. He made a distinction between the amount of information being stored and the rate of information flowing through a channel as *information flow* measured in *bits/sec* [17]. His theory is based on psychological experiments which suggested that conscious working memory can not take more than 16 *bits/sec* of visual information. Then he argued that artists should provide a flow of information of about 16 *bits/sec* for works of art to be perceived as beautiful and harmonious.

Staudek in his multi-criteria approach (informational and structural) as *exact aesthetics* to Birkhoff's measure applied information flow I' by defining it as a measure assessing principal information transmission qualities in time. He used $16\,bits/sec$ reference as channel capacity $C_r = 16\,bits/sec$ and a time reference of $8\,\mathrm{s}$ ($t_r = 8\,\mathrm{s}$) to argue that artefacts with $I > 128\,bits$ will not fit into the conscious working memory for absorbing the whole aesthetic message [37].

Machado and Cardoso proposed a model based on Birkhoff's approach as the ratio of *image complexity* to *processing complexity* by arguing that images with high visual complexity, are processed easily so they have highest aesthetic value [26]. Adapting Bense's informational aesthetics to different approaches of the concepts of order and complexity in an image, three measures based on Kolmogorov complexity [25], Shannon entropy (for RGB channels) and Zurek's physical entropy [40] were introduced. Then the measures were applied to analyse aesthetic values of several paintings (Mondrian, Pollock, and van Gogh) [33, 34].

3   Spatial Complexity Measure

Despite the dominance of entropy as a measure of order and complexity, it fails to capture structural characteristics of 2D patterns. The main reason for this drawback is that entropy is a function of the distribution of the symbols, and not on their spatial arrangement [23]. Consequently any model derived from information theory will inherently suffer from this drawback.

Considering our intuitive perception of complexity and structural characteristics of 2D patterns, a complexity measure must be bounded by two extreme points of complete order and disorder. It is reasonable to assume that *regular structures*, *irregular structures* and *structureless* patterns lie along between these extremes, as illustrated in Fig. 1.

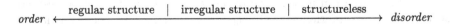

Fig. 1. The spectrum of spatial complexity.

A complete regular structure is a pattern of high symmetry, an irregular structure is a pattern with some sort of structure but not as regular as a fully symmetrical pattern and finally a structureless pattern is a random arrangement of elements [22]. A measure introduced in [1, 5, 38] and known as *information gain*, has been proposed as a means of characterising the complexity of dynamical systems and of 2D patterns. It measures the amount of information gained in bits when specifying the value, x, of a random variable X given knowledge of the value, y, of another random variable Y,

$$G_{x,y} = -\log_2 P(x|y). \tag{4}$$

$P(x|y)$ is the conditional probability of a state x conditioned on the state y. Then the *mean information gain* (MIG), $\overline{G}_{X,Y}$, is the average amount of information gain from the description of the all possible states of Y:

$$\overline{G}_{X,Y} = \sum_{x,y} P(x,y)G_{x,y} = -\sum_{x,y} P(x,y)\log_2 P(x|y) \tag{5}$$

where $P(x,y)$ is the joint probability, $\mathrm{prob}(X = x, Y = y)$. $\overline{G}$ is also known as the conditional entropy, $H(X|Y)$ [11]. Conditional entropy is the reduction in uncertainty of the joint distribution of X and Y given knowledge of Y, $H(X|Y) = H(X,Y) - H(Y)$. The lower and upper bounds of $\overline{G}_{X,Y}$ are

$$0 \leqslant \overline{G}_{X,Y} \leqslant \log_2 |\mathcal{X}|. \tag{6}$$

The structural characteristics of a 2D image are determined by the spatial distribution and state (i.e. colour) of individual pixels. In terms of the state of pixels they can be either with a uniform state relation (same colours) or non-uniform state relation (different colours) with their neighbouring pixels. In order to apply G to an image, the following definitions are needed:

- L is a finite lattice of pixels (i,j).
- $S = \{1, 2, \ldots, k\}$ is set of states. Each pixels (i,j) in L has a state $s \in S$.
- N is neighbourhood, as specified by a set of lattice vectors $\{e_a\}$, $a = 1, 2, \ldots, N$. The neighbourhood of pixel $r = (i,j)$ is $\{r+e_1, r+e_2, \ldots, r+e_N\}$. With an economy of notation, the pixels in the neighbourhood of (i,j) can be numbered from 1 to N; the neighbourhood states of (i,j) can therefore be denoted $(s_1, s_2, \ldots, s_N)$. An eight-cell neighbourhood $\{(\pm1, 0), (0, \pm1), (\pm1, \pm1)\}$ is considered for a pixel's relation to its neighbouring pixels. The relative positions for non-edge pixels, since they do not have neighbouring pixels, is given by matrix M:

$$M = \begin{bmatrix} (i-1,j+1) & (i,j+1) & (i+1,j+1) \\ (i-1,j) & (i,j) & (i+1,j) \\ (i-1,j-1) & (i,j-1) & (i+1,j-1) \end{bmatrix}. \tag{7}$$

For an image, $\overline{G}$ can be calculated by considering the distribution of pixel colours over pairs of pixels r, s,

$$\overline{G}_{r,s} = -\sum_{s_r, s_s} P(s_r, s_s)\log_2 P(s_r, s_s) \tag{8}$$

where s_r, s_s are the states at r and s. Since $|S| = N$, $\overline{G}_{r,s}$ is a value in $[0, N]$ (details of calculations for a sample pattern are provided in appendix). The vertical, horizontal, primary diagonal ($\setminus$) and secondary diagonal ($\diagup$) neighbouring pairs provide eight $\overline{G}$s; $\overline{G}_{(i,j),(i-1,j+1)}$, $\overline{G}_{(i,j),(i,j+1)}$, $\overline{G}_{(i,j),(i+1,j+1)}$, $\overline{G}_{(i,j),(i-1,j)}$, $\overline{G}_{(i,j),(i+1,j)}$, $\overline{G}_{(i,j),(i-1,j-1)}$, $\overline{G}_{(i,j),(i,j-1)}$ and $\overline{G}_{(i,j),(i+1,j-1)}$. Correlations between pixels on opposing lattice edges are not considered. The result of this edge condition is that $G_{i+1,j}$ is not necessarily equal to $\overline{G}_{i-1,j}$.

Figure 2 illustrates the advantages of $\overline{G}$ over H in discriminating structurally different patterns where the elements are equally probable ($P(s_r, s_s) = \frac{1}{108}$). Figure 2a is completely symmetrical, Fig. 2b is partially structured and Fig. 2c is a structureless and random pattern. The calculations have been performed for each element of images having a uniform and non-uniform colours in their relative spatial positions for three possible colours ($S = \{lightgrey, grey, black\}$) along with $\overline{G}$, and $\mu(\overline{G})$, the mean of the eight directional G's. As is evident, H is identical for these structurally different patterns, however, $\overline{G}$ and $\mu(\overline{G})$ reflect the order and complexity of patterns due to the spatial arrangements of composing elements. Figure 2 clearly demonstrates the drawbacks of entropy to discriminate structurally different 2D patterns. In other words, entropy is invariant to the spatial arrangement of the composing elements.

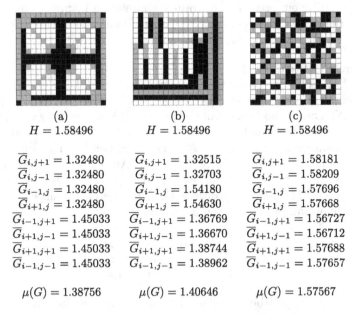

(a)	(b)	(c)
$H = 1.58496$	$H = 1.58496$	$H = 1.58496$
$\overline{G}_{i,j+1} = 1.32480$	$\overline{G}_{i,j+1} = 1.32515$	$\overline{G}_{i,j+1} = 1.58181$
$\overline{G}_{i,j-1} = 1.32480$	$\overline{G}_{i,j-1} = 1.32703$	$\overline{G}_{i,j-1} = 1.58209$
$\overline{G}_{i-1,j} = 1.32480$	$\overline{G}_{i-1,j} = 1.54180$	$\overline{G}_{i-1,j} = 1.57696$
$\overline{G}_{i+1,j} = 1.32480$	$\overline{G}_{i+1,j} = 1.54630$	$\overline{G}_{i+1,j} = 1.57668$
$\overline{G}_{i-1,j+1} = 1.45033$	$\overline{G}_{i-1,j+1} = 1.36769$	$\overline{G}_{i-1,j+1} = 1.56727$
$\overline{G}_{i+1,j-1} = 1.45033$	$\overline{G}_{i+1,j-1} = 1.36670$	$\overline{G}_{i+1,j-1} = 1.56712$
$\overline{G}_{i+1,j+1} = 1.45033$	$\overline{G}_{i+1,j+1} = 1.38744$	$\overline{G}_{i+1,j+1} = 1.57688$
$\overline{G}_{i-1,j-1} = 1.45033$	$\overline{G}_{i-1,j-1} = 1.38962$	$\overline{G}_{i-1,j-1} = 1.57657$
$\mu(G) = 1.38756$	$\mu(G) = 1.40646$	$\mu(G) = 1.57567$

Fig. 2. Measures of H, $\overline{G}$s and $\mu(G)$ for structurally different patterns with equally probable distribution of elements.

4   Analysis

The purpose of this section compare $\mu(G)$ for twelve patterns with an empirical aesthetic ranking.

Twelve experimental stimuli were adapted from an empirical study of human aesthetic judgement [21]. Jacobsen [21] reports on an empirical trial of human aesthetic judgement. Fifty-five young adults (15 males) participated in the experiment for course credit or partial fulfilment of course requirements. All were first or second year psychology students at the University of Leipzig. None of them had received professional training in the fine arts or participated in a similar experiment before. Participants reported normal or corrected-to-normal visual acuity. Subjects were asked to evaluate images from two groups; a group of asymmetrical images and a group of symmetrical images with at least one axis

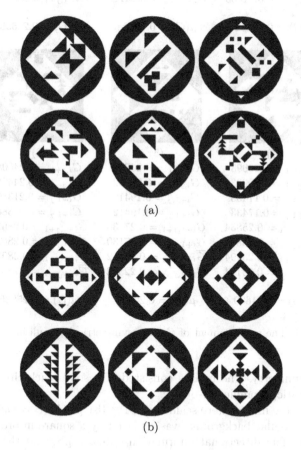

Fig. 3. Stimulus examples from [20]. The patterns in (a) are not symmetric, ranging from not beautiful (as judged in the trials) to beautiful (from top left for bottom right) and the patterns in (b) are symmetric, ranging also from not beautiful to beautiful (from top left for bottom right)

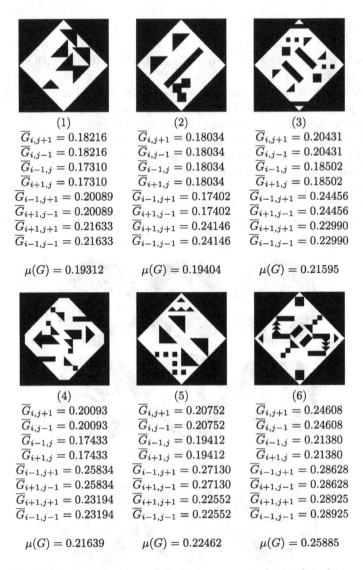

Fig. 4. The measurement of Gs for asymmetrical stimuli in *bits*.

of reflection symmetry. The images consisted of a solid black circle showing a centred, quadratic, rhombic cut-out.

Twelve of these images were scaled to 151×151 pixels ($S = \{white, black\}$) and the black circular background was replaced by a square in order to reduce aliasing errors. The directional information gains, Eq. 8, and the mean gain, $\mu(G)$, were calculated. The results are detailed in Figs. 4 and 6.

The relationship between aesthetic judgements and $\mu(G)$ for asymmetrical stimuli are shown in Fig. 5. The analysis shows a strong positive correlation

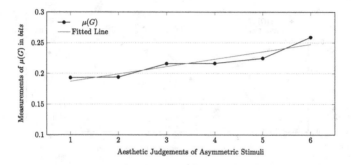

Fig. 5. The measurements of $\mu(G)$ for asymmetrical stimuli.

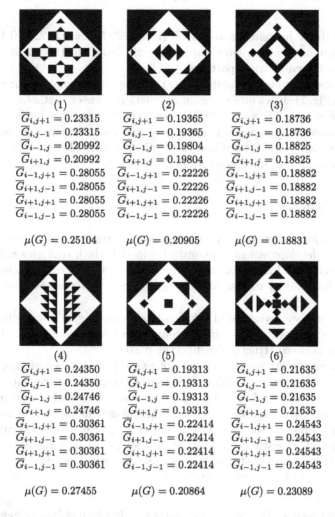

Fig. 6. The measurement of Gs for symmetrical stimuli in *bits*.

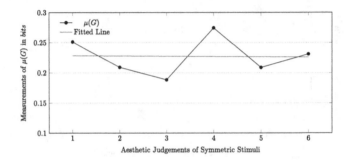

Fig. 7. The measurements of $\mu(G)$ for symmetrical stimuli.

between aesthetic judgements and $\mu(G)$ ($r = 0.9327$, $y = 0.012x + 0.175$). This indicates that information gain can be directly linked to the human aesthetic judgements of asymmetrical patterns.

The Pearson correlation coefficient (r) and regression analysis for symmetrical stimuli (Fig. 7) shows no significant correlation between aesthetic judgements and $\mu(G)$ ($r = -0.0266$, $y = 0.229$). However the images differ in their degree of reflection symmetry. Images 1, 2 and 3 have two axes of symmetry, image 4 has a single axes and 5 and 6 have four axes. This mixture of degrees of symmetry might account for the ambiguous results.

5   Conclusions

One of the major challenges of evolutionary art and computational notions of aesthetics is the development of a quantitative model which conforms with human intuitive perception. Informational theories of aesthetics based the measurements of entropy have failed to discriminate structurally different patterns in a 2D plane.

In this work, we investigated information gain as a spatial complexity measure. This measure which takes into account correlations between pixels and can discriminate between structurally different patterns.

This paper reports on the analysis of two different types of stimuli (symmetrical and asymmetrical) which were adapted from an experimental study on human aesthetic perception in the visual domain. The analysis suggest a link between information gain and aesthetic adjustments, in the case of asymmetrical patterns. However, the analysis did not show and link between information gain and empirical aesthetic judgement in the case of patterns with reflection symmetry. It is conjectured that having different orders of reflection symmetry has contributed to this negative finding.

Acknowledgements. We are grateful to Thomas Jacobsen of Helmut Schmidt University for granting permission to use his experimental stimuli.

Appendix

In this example the pattern is composed of two different colours $S = \{white, black\}$ where the set of permutations with repetition is $\{ww, wb, bb, bw\}$. Considering the mean information gain (Eq. 8) and given the matrix M (Eq. 7), the calculations can be performed as follows:

$white - white$

$P(w, w_{(i,j+1)}) = \frac{5}{6}$

$P(w|w_{(i,j+1)}) = \frac{4}{5}$

$P(w, w_{(i,j+1)}) = \frac{5}{6} \times \frac{4}{5} = \frac{2}{3}$

$G(w, w_{(i,j+1)}) = \frac{2}{3} \log_2 P(\frac{4}{5})$

$G(w, w_{(i,j+1)}) = 0.2146 \ bits$

$white - black$

$P(w, b_{(i,j+1)}) = \frac{5}{6}$

$P(w|b_{(j+1)}) = \frac{1}{5}$

$P(w, b_{(i,j+1)}) = \frac{5}{6} \times \frac{1}{5} = \frac{1}{6}$

$\overline{G}(w, b_{(i,j+1)}) = \frac{1}{6} \log_2 P\frac{1}{5}$

$\overline{G}(w, b_{(i,j+1)}) = 0.3869 \ bits$

$black - black$

$P(b, b_{(i,j+1)}) = \frac{1}{6}$

$P(b|b_{(i,j+1)}) = \frac{1}{1}$

$P(b, b_{(i,j+1)}) = \frac{1}{6} \times \frac{1}{1} = \frac{1}{6}$

$G(b, b_{(i,j+1)}) = \frac{1}{6} \log_2 P(1)$

$G(b, b_{(i,j+1)}) = 0 \ bits$

$black - white$

$P(b, b_{(i,j+1)}) = \frac{1}{6}$

$P(b|w_{(i,j+1)}) = \frac{0}{1}$

$P(b, w_{(i,j+1)}) = \frac{1}{6} \times 0$

$G(b, w_{(i,j+1)}) = 0 \ bits$

$\overline{G} = G(w, w_{(i,j+1)}) + G(w, b_{(i,j+1)}) + G(b, b_{(i,j+1)}) + G(b, w_{(i,j+1)})$

$\overline{G} = 0.6016 \ bits$

In $white - white$ case G measures the uniformity and spatial property where $P(w, w_{(i,j+1)})$ is the joint probability that a pixel is white and it has a neighbouring pixel at its $(i, j+1)$ position, $P(w|w_{(i,j+1)})$ is the conditional probability of a pixel is white given that it has white neighbouring pixel at its $(i, j+1)$ position, $P(w, w_{(i,j+1)})$ is the joint probability that a pixel is white and it has neighbouring pixel at its $(i, j+1)$ position, $G(w, w_{(i,j+1)})$ is information gain in $bits$ from specifying a white pixel where it has a white neighbouring pixel at its $(i, j+1)$ position. The same calculations are performed for the rest of cases; $black$-$black$, $white$-$black$ and $black$-$white$.

References

1. Andrienko, Y.A., Brilliantov, N.V., Kurths, J.: Complexity of two-dimensional patterns. Eur. Phys. J. B **15**(3), 539–546 (2000)
2. Arnheim, R.: Art and Visual Perception: A Psychology of the Creative Eye. Univ of California Press, Berkeley (1954)
3. Arnheim, R.: Towards a Psychology of Art/entropy and Art an Essay on Disorder and Order. The Regents of the University of California (1966)
4. Arnheim, R.: Visual Thinking. Univ of California Press, Berkeley (1969)
5. Bates, J.E., Shepard, H.K.: Measuring complexity using information fluctuation. Phys. Lett. A **172**(6), 416–425 (1993)

6. Bense, M., Nee, G.: Computer grafik. In: Bense, M., Walther, E. (eds.) Edition Rot, vol. 19. Walther, Stuttgart (1965)

7. Bense, M.: Aestetica: Programmierung des Schönen, allgemeine Texttheorie und Textästhetik [Aesthetica : Programming of beauty, general text theory and aesthetics]. Agis-Verlag (1960)

8. Bense, M.: Kleine abstrakte ästhetik [small abstract aesthetics]. In: Walther, E. (ed.) Edition Rot, vol. 38 (1969)

9. Birkhoff, G.: Aesthetic Measure. Harvard University Press, Cambridge (1933)

10. Ciesielski, V., Barile, P., Trist, K.: Finding image features associated with high aesthetic value by machine learning. In: Machado, P., McDermott, J., Carballal, A. (eds.) EvoMUSART 2013. LNCS, vol. 7834, pp. 47–58. Springer, Heidelberg (2013)

11. Cover, T.M., Thomas, J.A.: Elements of Information Theory. Wiley Series in Telecommunications and Signal Processing. Wiley-Interscience, New York (2006)

12. Dawkins, R.: The Blind Watchmaker. W. W. Norton, New York (1986)

13. den Heijer, E., Eiben, A.E.: Comparing aesthetic measures for evolutionary art. In: Chio, C., Brabazon, A., Caro, G.A., Ebner, M., Farooq, M., Fink, A., Grahl, J., Greenfield, G., Machado, P., O'Neill, M., Tarantino, E., Urquhart, N. (eds.) EvoApplications 2010, Part II. LNCS, vol. 6025, pp. 311–320. Springer, Heidelberg (2010)

14. Eysenck, H.J.: An experimental study of aesthetic preference for polygonal figures. J. Gen. Psychol. **79**(1), 3–17 (1968)

15. Eysenck, H.J.: The empirical determination of an aesthetic formula. Psychol. Rev. **48**(1), 83 (1941)

16. Eysenck, H.J.: The experimental study of the 'good gestalt' –a new approach. Psychol. Rev. **49**(4), 344 (1942)

17. Franke, H.W.: A cybernetic approach to aesthetics. Leonardo **10**(3), 203–206 (1977)

18. Galanter, P.: Computational aesthetic evaluation: past and future. In: McCormack, J., d'IInverno, M. (eds.) Computer and Creativity, pp. 255–293. Springer, Heidelberg (2012)

19. den Heijer, E.: Autonomous Evolutionary Art, Ph.D. thesis. Vrije Universiteit, Amsterdam (2013)

20. Jacobsen, T.: Beauty and the brain: culture, history and individual differences in aesthetic appreciation. J. Anat. **216**(2), 184–191 (2010)

21. Jacobsen, T., Hofel, L.: Aesthetic judgments of novel graphic patterns: analyses of individual judgments. Percept. Mot. Skills **95**(3), 755–766 (2002)

22. Javaheri Javid, M.A., Blackwell, T., Zimmer, R., Al-Rifaie, M.M.: Spatial complexity measure for characterising cellular automata generated 2D patterns. In: Pereira, F., Machado, P., Costa, E., Cardoso, A. (eds.) EPIA 2015. LNCS, vol. 9273, pp. 201–212. Springer, Heidelberg (2015)

23. Javid, M.A.J., al-Rifaie, M.M., Zimmer, R.: An informational model for cellular automata aesthetic measure. In: AISB Symposium on Computational Creativity. University of Kent, Canterbury, UK (2015)

24. Latham, W.H., Todd, S.: Computer sculpture. IBM Syst. J. **28**(4), 682–688 (1989)

25. Li, M.: An introduction to Kolmogorov complexity and its applications. Springer, New York (1997)

26. Machado, P., Cardoso, A.: Computing aesthetics. In: de Oliveira, F.M. (ed.) SBIA 1998. LNCS (LNAI), vol. 1515, pp. 219–228. Springer, Heidelberg (1998)

27. Machado, P., Romero, J., Manaris, B.: Experiments in computational aesthetics: an iterative approach to stylistic change in evolutionary art. In: Romero, J., Machado, P. (eds.) The Art of Artificial Evolution: A Handbook on Evolutionary Art and Music, pp. 381–415. Springer, Heidelberg (2008)

28. McCormack, J.: Open problems in evolutionary music and art. In: Rothlauf, F., Branke, J., Cagnoni, S., Corne, D.W., Drechsler, R., Jin, Y., Machado, P., Marchiori, E., Romero, J., Smith, G.D., Squillero, G. (eds.) EvoWorkshops 2005. LNCS, vol. 3449, pp. 428–436. Springer, Heidelberg (2005)

29. McCormack, J.: Facing the future: evolutionary possibilities for human-machine creativity. In: Romero, J., Machado, P. (eds.) The Art of Artificial Evolution, pp. 417–451. Springer, Heidleberg (2008)

30. Moles, A.: Information Theory and Esthetic Perception. Illinois Press, Urbana (1968). Trans. JE Cohen. U

31. Nake, F.: Information aesthetics: an heroic experiment. J. Math. Arts 6(2–3), 65–75 (2012)

32. Noll, A.M.: The digital computer as a creative medium. IEEE Spectr. 4(10), 89–95 (1967)

33. Rigau, J., Feixas, M., Sbert, M.: Informational aesthetics measures. IEEE Comput. Graph. Appl. 28(2), 24–34 (2008)

34. Rigau, J., Feixas, M., Sbert, M.: Conceptualizing Birkhoff's aesthetic measureusing shannon entropy and kolmogorov complexity. In: Cunningham, D.W., Meyer, G., Neumann, L. (eds.) Workshop on Computational Aesthetics, pp. 105–112. Eurographics Association, Banff, Alberta, Canada (2007)

35. Shannon, C.: A mathematical theory of communication. Bell Syst. Tech. J. 27, 379–423, 623–656 (1948)

36. Sims, K.: Artificial evolution for computer graphics. Technical Report, TR-185, Thinking Machines Corporation (1991)

37. Staudek, T.: Exact aesthetics, object and scene to message. Ph.D. thesis, Faculty of Informatics, Masaryk University of Brno (2002)

38. Wackerbauer, R., Witt, A., Atmanspacher, H., Kurths, J., Scheingraber, H.: A comparative classification of complexity measures. Chaos, Solitons & Fractals 4(1), 133–173 (1994)

39. Wilson, D.J.: An experimental investigation of Birkhoff's aesthetic measure. J. Abnorm. Soc. Psychol. 34(3), 390 (1939)

40. Zurek, W.H.: Algorithmic randomness and physical entropy. Phys. Rev. A 40(8), 4731 (1989)

Exploring the Visual Styles
of Arcade Game Assets

Antonios Liapis[(✉)]

Institute of Digital Games, University of Malta, Msida, Malta
antonios.liapis@um.edu.mt

Abstract. This paper describes a method for evolving assets for video games based on their visuals properties. Focusing on assets for a space shooter game, a genotype consisting of turtle commands is transformed into a spaceship image composed of human-authored sprite components. Due to constraints on the final spaceships' plausibility, the paper investigates two-population constrained optimization and constrained novelty search methods. A sample of visual styles is tested, each a combination of visual metrics which primarily evaluate balance and shape complexity. Experiments with constrained optimization of a visual style demonstrate that a visually consistent set of spaceships can be generated, while experiments with constrained novelty search demonstrate that several distinct visual styles can be discovered by exploring along select, or all, visual dimensions.

1 Introduction

The design and development of digital games requires an extraordinary amount of creativity from the part of human game developers. Digital games rely on several creative domains such as visual art, sound design, interaction design, narrative, virtual cinematography, aesthetics and environment beautification [1]. All of these creative domains are fused together in a single software application, which must be engaging and entertaining to a broad range of players with diverse tastes and skillsets. While game development still remains primarily a human endeavor, there has been considerable interest both within the game industry and within academia in the automatic, algorithmic generation of game elements. Procedural content generation for games serves two purposes from the perspective of commercial game developers: (a) it alleviates much of the effort of manual asset creation as games are "going to need a lot more content" in the future [2], (b) it can increase the replayability value of the game as players constantly encounter new content in a "perpetually fresh world" [3]. While most commercial and academic applications of procedural content generation revolve around game levels or rules and evaluate their playability and tractable properties [4], there is merit in considering the visual appeal of the generated content and in developing computational creators that can exhibit creativity comparable to that of human game visual artists.

© Springer International Publishing Switzerland 2016
C. Johnson et al. (Eds.): EvoMUSART 2016, LNCS 9596, pp. 92–109, 2016.
DOI: 10.1007/978-3-319-31008-4_7

This paper describes a generator of game assets grounded on visual principles of human perception. Specifically, constrained evolution creates spaceship sprites which can be used as player-controlled avatars or waves of enemies for an arcade space shooter game such as *Ikaruga* (Treasure 2008) and *M.U.S.H.A.* (Compile 1990). Towards that end, several constraints on the plausibility of spaceships are encoded, and their satisfaction guaranteed via two-population constrained evolution. Spaceships which satisfy the constraints are evaluated on certain visual properties, mostly revolving around balance and shape complexity [5]. These visual metrics are aggregated into objective functions which are optimized by a feasible-infeasible two-population genetic algorithm [6]; results show that evolved content can exhibit a consistent visual style as specified by a designer who combines desirable metrics into the algorithm's objective. Moreover, the same visual metrics are used to measure the difference between generated spaceships in a feasible-infeasible novelty search algorithm [7]; results show that the exploration afforded by novelty search discovers several distinct visual styles in an unsupervised fashion. The initial investigation in this paper can be expanded by including constraints and evaluations on the spaceships' in-game behavior, by expanding the range of visual metrics considered or by deriving visual quality from machine-learned models.

This paper contributes to the literature on computational game creativity [1] by fulfilling a graphic design task via evolutionary algorithms. More importantly, the experiments test how novelty search performs in constrained spaces as feasible-infeasible novelty search, and compare the impact of different distance functions on the appearance and variety of evolved assets. This constitutes a substantial contribution to constrained novelty search, which is still an under-explored topic [7,8], and more generally to the study on novelty search. Finally, the analysis regarding both the quality and the diversity of the final evolved content is a contribution to search-based procedural content generation research, especially concerning generators' expressivity [9,10].

2   Related Work

The evolutionary algorithms used for generating spaceships fall under the domain of constrained optimization and constrained novelty search, while the constraints and evaluations of generated content originate from a broad literature on search-based procedural content generation for games. This section highlights work on these related domains.

2.1   Procedural Content Generation

Using algorithmic methods to generate content for games has a long history within the game industry, dating back to *Rogue* (Toy and Wichman, 1980) and *Elite* (Acornsoft, 1984) which used procedural content generation (PCG) to create the game's dungeons and galaxies respectively. Currently, PCG is often used in games created by small teams in order to quickly create vast gameworlds such as those in *No Man's Sky* (Hello Games 2016) and engage players with perpetually fresh content such as new dungeons in *Torchlight* (Runic 2009) or new noble bloodlines in *Crusader Kings II* (Paradox 2012).

Academic interest in PCG for games is relatively recent but has received significant attention in the last decade, particularly for the application of artificial evolution for PCG under the umbrella term of *search-based PCG* [11]. Search-based PCG often requires carefully designed objective functions, which are often inspired by game design theory and design patterns [12], and may target player balance in multi-player games [13] or challenge in single player games [14]. Since generated content often come with playability concerns, constrained evolution has been applied to the generation of traversable game levels [14], as well as in generators used as part of the design process [15]. However, the types of content generated in search-based PCG and the objective functions of such projects are surprisingly narrow in scope: most generators attempt to optimize game levels of some sort, while the fitness functions evaluate purely functional properties such as safety of resources and reachability of remote locations. Few search-based PCG projects aim to generate graphical assets, or evaluate them on their visual qualities: of note are the game shaders of [16] which evolved towards a designer-specified color, the weapon particles of [17] which evolved based on users' gameplay, and flowers of [18] which evolved via playful interactive evolution. Previous work by the author on spaceship generation [19] revolved around evolving the spaceships' geometry directly, and used fitness functions inspired by visual perception [5]. The current paper is motivated both by these earlier findings and by the gap in search-based PCG literature regarding graphical asset generation and visual quality evaluation.

2.2   Constrained Optimization and Constrained Novelty Search

Artificial evolution has traditionally been challenged in constrained spaces, as the division of evolving phenotypes between *feasible* (i.e. satisfying all constraints) and *infeasible* (i.e. failing one or more constraints) does not easily provide a gradient for search. A naive approach is to assign infeasible individuals the lowest fitness (*death penalty*), which has been argued against as infeasible individuals' valuable genetic information is lost [20]. Instead, a popular solution is to apply a penalty to the fitness of infeasible individuals [21]. Designing penalty functions can be challenging as a high penalty can become a death penalty while a low penalty can lead to extraneous exploration of infeasible space. Instead, [6] suggest that evolving infeasible individuals in a separate population to feasible ones circumvents the need to compare feasible with infeasible individuals. The suggested feasible-infeasible two-population genetic algorithm (FI-2pop GA) evolves an infeasible population towards minimizing the distance to feasibility while the feasible population evolves towards the problem-specific objective. Bringing infeasible individuals closer to the feasibility border increases the chances that they will produce feasible offspring; feasible offspring of infeasible individuals migrate to the feasible population and vice versa. The FI-2pop GA has been applied to search-based PCG for games, generating spaceships [22] and game levels [12,14].

Novelty search [23] has recently been proposed as an alternative to objective-driven search. Novelty search prioritizes diversifying the population rather than gradually improving an objective score, and can outperform objective-driven search when the fitness function is deceptive, subjective, or unknown. Novelty search selects individuals based on their average distance to their nearest neighbors; these neighbors can be in the current population or in an archive of novel individuals. In each generation, novelty search may store the population's most novel individuals in this archive, which acts as a form of memory and promotes exploration of yet unvisited areas of the search space.

As with objective-driven search, novelty search can suffer from a search space divided into feasible and infeasible areas. Minimal-criteria novelty search applies the death penalty (a fitness of 0) to infeasible individuals [8], but can suffer from lost genetic information. Using a two-population approach, feasible-infeasible novelty search (FINS) evolves the feasible population according to novelty search, diversifying feasible individuals from the current population and from an archive of feasible individuals [7]. Since the two-population approach can apply different selection criteria on each population, the feasible population diversifies its members via novelty search, while the infeasible population minimizes its members' distance from feasibility. Moreover, the feasible population creates an archive of novel individuals (containing only feasible results) to better explore the feasible search space. Feasible-infeasible dual novelty search (FI2NS) evolves both the feasible and the infeasible population towards novelty [7], and uses a different novel archive (with feasible and infeasible individuals respectively) for each population. Experiments with FINS and FI2NS in [24] have shown that FINS can quickly and reliably discover feasible individuals in highly constrained spaces, while FI2NS can create more diverse results in less constrained spaces.

3 Methodology

This paper uses constrained evolutionary approaches to generate spaceships constructed from multiple human-authored sprite components: the genetic encoding of these spaceships is detailed in Sect. 3.1. Section 3.2 describes the evolutionary algorithms, genetic operators and selection strategies used. Finally, Sect. 3.3 provides an overview of the constraints for plausible spaceships and the metrics for assessing visual style.

3.1 Spaceship Representation

The spaceships generated by the evolutionary algorithms are produced via turtle graphics, a drawing method where a cursor (turtle) moves upon a Cartesian plane. In this case, every step of the turtle equates to the placement of an image (i.e. a *sprite component*) taken from a library of human-authored images. The type of image and its filename is specified in the turtle command: the library

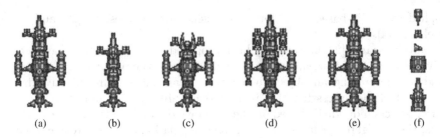

Fig. 1. An example spaceship and its mutated offspring. The spaceship in (a) is constructed from the schema `UT15[RN01]UB12UB11[RN01RB08[UW02]DT02]UB05UC12 [RW00]` which forms its genotype. The sprite itself is constructed via a turtle which places human-authored component sprites, a sample of which is shown in (f) (from top to bottom: `T15`, `W00`, `N01`, `B11` and `C12`). The remaining spaceships are produced from a single mutation of the spaceship in (a). (b) applies the remove mutation; (c) applies the change sprite mutation. (d) applies the expand vertically mutation under the weapon attachments of the cockpit; (e) applies the expand horizontally mutation, as side sponsons left and right of the tail fins.

contains multiple weapon sprites (`W`), thruster sprites (`T`), body sprites (`B`), cockpit sprites (`C`), and connector sprites (`N`); see Fig. 1(f) for examples of each type. Each turtle command also specifies whether the sprite will be placed to the right (`R`), left (`L`), upwards (`U`) or downwards (`D`) of the current sprite. Each spaceship is thus encoded in multiple commands such as `RT01`, i.e. attach the thruster sprite (`T`) with ID `01` to the right of the current sprite. In order to allow for more complex, branching shapes, the `[` command pushes the current position and sprite of the cursor to a stack, while the `]` command restores the cursor to the previous position and sprite on the stack (removing it from the stack in the process).

The sequence of turtle commands which encode a spaceship is called a *schema*, and acts as the genotype for the evolutionary process. This schema is translated into a fully colored spaceship (see Fig. 1(a)); however, several constraints and fitness evaluations operate on the spaceship's *footprint* which treats all pixels of the spaceship as one color. The first command of a schema is placed at the center of a canvas: in this paper, the canvas has a preset size of 200 by 360 pixels. The schema describes only the right half of the spaceship: after running all the turtle's commands, spaceships are rendered symmetrical by reflecting the sprites horizontally, along the canvas center.

3.2   Spaceship Evolution

The genotype of every spaceship is an array of characters which act as commands (or parts of commands) for the turtle graphics. These genotypes are evolved via mutation alone: the mutation operators are chosen carefully in order to minimize the likelihood of infeasible or undesirable spaceships being created. An offspring is generated by iteratively applying mutation operators 1 to 4 times (chosen randomly) on the selected parent. There are four types of mutation possible in this evolutionary algorithm, as shown in Fig. 1(b)–(e): a mutation can remove one or more turtle commands, change the sprite ID of a turtle command, and expand a schema vertically or horizontally.

This paper tests spaceship evolution using constrained optimization (FI-2pop GA) and constrained novelty search (FINS) methods. Both methods use a two-population approach [6], where individuals which satisfy all constraints evolve in a different population than individuals which fail one or more constraints. In both FINS and FI-2pop GA, individuals in the infeasible population use objective-driven evolution, attempting to minimize their distance from feasibility which is aggregated from several constraints described in Sect. 3.3. The feasible population in FI-2pop GA evolves towards maximizing an objective pertaining to a specific visual style. For FINS, the feasible population evolves towards maximizing the novelty score ρ of Eq. (1), which evaluates the average distance of an individual to its closest neighbors in the current population and in an archive of past novel discoveries. In all experiments in this paper, the 15 closest neighbors are considered when calculating novelty ($k = 15$), while in every generation the most novel feasible individual is added to the novel archive, which is initially empty.

$$\rho(i) = \frac{1}{k} \sum_{j=1}^{k} d(i, \mu_j) \tag{1}$$

where μ_j is the j-th-nearest neighbor of i (within the population and in the archive of novel individuals); distance $d(i,j)$ is a domain-dependent metric.

In all algorithms in this paper, parents are chosen via fitness-proportionate roulette wheel selection. The same parent can be chosen more than once, thus producing multiple mutated offspring. The fittest individual is carried over to the next generation (elitism of 1). Finally, the two evolving populations are artificially balanced via the *feasible offspring boost*, which is applied when the feasible population is smaller than the infeasible one: this mechanism assigns an equal number of offspring to both feasible and infeasible population (i.e. half of the total population) regardless of the number of parents in each. Although it depends on the chance of infeasible offspring from feasible parents, the feasible offspring boost generally increases the feasible population's size.

3.3   Spaceship Evaluation

Generated spaceships must fulfill certain criteria in order to be plausible. Spaceships which satisfy such constraints evolve to exhibit certain visual properties, while spaceships which do not satisfy one or more constraints are evaluated based on their distance to feasibility. Spaceships must satisfy the following constraints in order to be feasible:

c_b Spaceships can not extend past the borders of the canvas (200 by 360 pixels).
c_s Spaceships must be a single segment (measured via a 4-directional flood fill).

c_o Spaceships can not have sprite components overlapping with other components.

c_w No sprite components can be in the weapons' line of fire.

c_t No sprite components except connectors can be in the thrusters' path.

c_m A spaceship's bounding box must not cover a surface smaller than 1024 pixels, as it will be too small for in-game use. Moreover, this constraint ensures that a feasible spaceship contains at least two sprite components.

Each of these constraints returns a score indicative of its distance from feasibility; only spaceships with a total distance of 0 for all constraints are feasible. Spaceships which satisfy all constraints are evaluated on their visual properties. Inspired by [5] and largely based on fitnesses implemented in [19], this paper evaluates primarily a spaceship's *balance*, operating either on its footprint or its bounding box. Thus, the spaceship is considered only in terms of its shape, ignoring its color; color may be considered in future work, however. This paper uses the following metrics:

f_t The ratio between top half and bottom half of the spaceship's footprint (see Fig. 2a).

f_{m_x} The ratio between the middle half of the spaceship's blueprint on the x axis (see Fig. 2b) and the remaining areas.

f_{m_y} The ratio between the middle half of the spaceship on the y axis (see Fig. 2c) and the remaining areas.

f_w The ratio between width and height of the spaceship's bounding box.

f_{bb} The ratio between the surface of the spaceship's footprint and the surface of its bounding box.

f_o The ratio between the spaceship's outline (see Fig. 2d) and the perimeter of its bounding box.

f_c The total number of pixels between adjacent components which do not match (see Fig. 2e), normalized to the total pixels on the borders of adjacent components.

Except for f_c, all other metrics are not bound and have a value range of $(0,\infty)$, with a score of 1 indicating a balanced ratio (e.g. a square bounding box for f_w). Therefore all scores except f_c are normalized via the sigmoid of Eq. (2) centered at $x = 1$. Due to the emphasis that $n(x)$ places on x values around 1 (e.g. slight imbalances between top and bottom halves in f_t), compared to small changes in $n(x)$ for large values of x, the normalized metrics are ideal for aggregating into a weighted sum for evolving spaceships towards multiple objectives of visual quality.

$$n(x) = \frac{1}{1 + e^{-5(x-1)}} \tag{2}$$

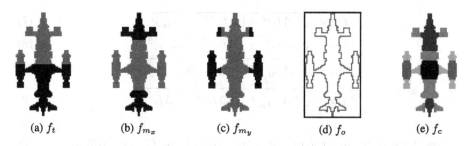

(a) f_t (b) f_{m_x} (c) f_{m_y} (d) f_o (e) f_c

Fig. 2. Evaluations of visual properties for the spaceship in Fig. 1(a): f_t evaluates the surface ratio between the top half (red) and the bottom half (black); f_{m_x} evaluates the surface between the middle half of the spaceship, split horizontally (red), with the remaining areas (black); f_{m_y} evaluates the surface between the middle half of the spaceship, split vertically (red), with the remaining areas (black); f_o evaluates the ratio between the length of the spaceship's outline (red) and the perimeter of the bounding box (black frame); f_o evaluates whether connected sprite components (in different shades of gray) have imperfect connections (in red) (Color figure online).

4   Experiments

In order to assess how evolution can achieve different visual styles in generated spaceships, the following indicative *style objective functions* are considered:

$$F_B = f_t + f_w + (1 - f_c) \tag{3}$$

$$F_O = f_o + f_{bb} + (1 - f_c) \tag{4}$$

$$F_H = f_{m_y} + f_{m_x} + (1 - f_c) \tag{5}$$

$$F_S = (1 - f_{m_y}) + (1 - f_t) + (1 - f_c) \tag{6}$$

Each of these objective functions describes a distinct visual style: F_B favors wide, front-loaded spaceships; F_O favors bulky, boxy spaceships (due to f_{bb}) which however have an interesting, complex outline (due to f_o); F_H favors spaceships which are aligned vertically and horizontally along the spaceship's centerpoint; finally, F_S favors spaceships with large side sponsons, while most of the main hull is located in the back. Note that all style objective functions attempt to minimize f_c; this fitness component is not a visual objective per se, but rather a soft constraint in order to deter tangential connections between sprite components which could lessen the plausibility of the spaceship.

While the functions of Eqs. (3)–(6) can be used as explicit objectives for a FI-2pop GA, novelty search requires a distance function to calculate its novelty score in Eq. (1). Distance functions can be derived directly from the objectives of Eqs. (3)–(6) and target diversity along those specific visual properties. These *style distance functions* are shown in Eqs. (7)–(10), where Δf is the difference in scores of two spaceships for metric f.

$$D_B = \sqrt{(\Delta f_t)^2 + (\Delta f_w)^2 + (\Delta f_c)^2} \tag{7}$$

$$D_O = \sqrt{(\Delta f_o)^2 + (\Delta f_{bb})^2 + (\Delta f_c)^2} \tag{8}$$

$$D_H = \sqrt{(\Delta f_{m_y})^2 + (\Delta f_{m_x})^2 + (\Delta f_c)^2} \tag{9}$$

$$D_S = \sqrt{(\Delta f_{m_y})^2 + (\Delta f_t)^2 + (\Delta f_c)^2} \tag{10}$$

Novelty search using the distance functions of Eqs. (7)–(10) will likely result in different visual styles along those visual dimensions. However, evaluating difference in all visual metrics of Sect. 3.3 can prompt novelty search to explore a broader spectrum of possible spaceship designs. Using the distance function of Eq. (11), novelty search can distinguish in a more granular fashion the visual differences between two spaceships, allowing it to simultaneously explore all combinations of visual metrics.

$$D_{all} = \sqrt{(\Delta f_t)^2 + (\Delta f_{m_x})^2 + (\Delta f_{m_y})^2 + (\Delta f_w)^2 + (\Delta f_{bb})^2 + (\Delta f_o)^2 + (\Delta f_c)^2} \tag{11}$$

In the experiments documented in the next section, four different FI-2pop GAs use the style objective functions of Eqs. (3)–(6) as their feasible fitness. Moreover, four different FINS algorithms use the distance functions of Eqs. (7)–(10) for novelty search on the feasible population; finally, a FINS algorithm identified as FINS$_{all}$ uses D_{all} of Eq. (11) to diversify its feasible individuals. For all algorithms, the infeasible population minimizes the distance from feasibility for all constraints outlined in Sect. 3.3. Results are collected from 50 evolutionary runs, after 100 generations with a total population of 100 individuals per run (including feasible and infeasible ones). Significance reported in this paper uses a two-tailed t-test assuming unequal variance, with a significance level $\alpha = 0.05$; when performing multiple comparisons, the Bonferroni correction [25] is applied. In reported t-tests, normality is established via the Shapiro-Wilk test [26].

4.1   Quality of Generated Results

In order to evaluate the quality of generated spaceships, the fittest individual at the end of each evolutionary run is found using the style objective functions of Eqs. (3)–(6). The score of these individuals in the style objective and its visual metrics are shown in Table 1. While for FINS$_{all}$ results in the table are collected from the same set of 50 evolutionary runs, all other algorithms explicitly use the same objective and distance scores as the style objective which determines the best individuals (i.e. the best spaceships for F_O in Table 1 were collected from runs targeting F_O in FI-2pop GA and D_O in FINS).

Observing the results of Table 1, it is clear that for all visual styles explored in Eqs. (3)–(6) the objective-driven FI-2pop GA attains higher fitness scores than the constrained novelty search variants. In all cases the FI-2pop GA attains a significantly higher style objective score in its best individuals than both FINS

Table 1. Feasible individuals and best individuals' fitness scores at the end of each evolutionary run. Results are averaged from 50 independent runs with standard deviation in parentheses.

F_B					
Method	Feasible	F_B	f_w	f_t	$1 - f_c$
FINS_{all}	41.3 (3.8)	0.89 (0.06)	0.94 (0.10)	0.97 (0.07)	0.77 (0.10)
FINS	44.3 (4.3)	0.87 (0.06)	0.89 (0.15)	0.94 (0.07)	0.76 (0.13)
FI-2pop GA	44.4 (4.3)	0.99 (0.01)	1.00 (0.01)	1.00 (0.00)	0.98 (0.03)
F_O					
Method	Feasible	F_O	f_o	f_{bb}	$1 - f_c$
FINS_{all}	41.3 (3.8)	0.80 (0.04)	0.92 (0.08)	0.66 (0.11)	0.82 (0.07)
FINS	42.9 (4.4)	0.79 (0.04)	0.85 (0.16)	0.68 (0.15)	0.84 (0.08)
FI-2pop GA	42.9 (5.2)	0.93 (0.01)	0.97 (0.02)	0.88 (0.04)	0.95 (0.04)
F_H					
Method	Feasible	F_H	f_{m_y}	f_{m_x}	$1 - f_c$
FINS_{all}	41.3 (3.8)	0.94 (0.02)	0.99 (0.01)	0.99 (0.01)	0.83 (0.06)
FINS	41.4 (4.0)	0.92 (0.03)	0.96 (0.06)	0.98 (0.04)	0.82 (0.08)
FI-2pop GA	46.8 (2.8)	0.99 (0.01)	1.00 (0.01)	1.00 (0.00)	0.98 (0.02)
F_S					
Method	Feasible	F_S	$1 - f_{m_y}$	$1 - f_t$	$1 - f_c$
FINS_{all}	41.3 (3.8)	0.80 (0.07)	0.76 (0.19)	0.88 (0.16)	0.77 (0.11)
FINS	43.0 (4.2)	0.81 (0.05)	0.82 (0.11)	0.85 (0.09)	0.77 (0.11)
FI-2pop GA	43.9 (3.9)	0.94 (0.02)	0.93 (0.04)	0.95 (0.02)	0.93 (0.05)

and FINS_{all}. This is not surprising, since FI-2pop GA explicitly rewards a specific visual style, and the same heuristic is then used to find the best individuals. While the FI-2pop GA could potentially under-perform if the fitness landscape was deceptive, this apparently is not the case for the representation and mutation operators used and the fitnesses tested.

A more surprising finding of Table 1 is that the general novelty search of FINS_{all} finds individuals of comparable (and in some cases higher) fitness as the more targeted FINS. Due to high deviation in both algorithms, only in experiments with F_B and F_H does FINS_{all} show a statistically significant increase from FINS. This behavior is likely due to the fact that FINS_{all} attempts to explore broadly, driven by a more granular distance function which differentiates spaceships based on many more visual properties than the style distance functions of Eqs. (7)–(10); through an indirect, serendipitous traversal of the search space it reaches areas of high scores in many visual metrics. How this exploration affects the diversity of results will be elaborated in Sect. 4.2.

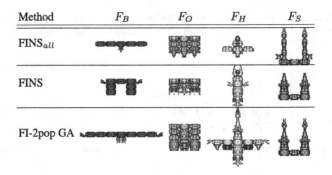

Fig. 3. Overall fittest spaceships in 50 independent runs per approach and fitness function.

Figure 3 shows the fittest individuals among the 50 independent runs for the different style objectives and algorithms. Different sprite colors are used to highlight the different style objectives. All evolutionary methods create spaceships which exhibit the intended visual properties of the aggregated fitness functions. For F_O, the best spaceships are boxy; for F_B they are wide and very elongated, and the bottom of the spaceship contains only thrusters; for F_H they are more "centrally" located around the image's center, with most spaceship components extending vertically and horizontally from that point; for F_S they have sizable side-sponsons and a smaller main hull at the bottom of the spaceship. While all style objectives reward a low f_c score, this is difficult to pinpoint in the best spaceships: as noted when defining the style objectives, minimizing f_c acts as a control mechanism rather than an explicit visual goal; only cases with high f_c would make its presence obvious as spaceships would appear almost disjointed.

4.2 Diversity of Generated Results

The diversity of generated results is evaluated in this paper from the perspective of visual properties identified in Sect. 3.3 and particularly the visual styles included in Eqs. (3)–(6), rather than a general visual difference. Ignoring the f_c dimension as it is a control mechanism with little visual impact, all the spaceships collected by the end of each evolutionary approach are evaluated on the remaining two dimensions of Eqs. (3)–(6) and plotted on Fig. 4. As with experiments in Sect. 4.1, spaceships from the same 50 evolutionary runs for FINS$_{all}$ are used for all plots, while spaceships for FINS and FI-2pop GAs explicitly use the same style distance and style objective as the metrics plotted (i.e. for the F_O plot, spaceships are collected from 50 independent runs of FINS using D_O as its distance function and FI-2pop GA using F_O as its objective).

Observing Fig. 4, most spaceships evolved via the FI-2pop GA unsurprisingly have high scores in both visual dimensions of the style objective function being optimized (with most points concentrated at the top right corner of each plot). Evolution towards each style objective creates a different spread, however, as

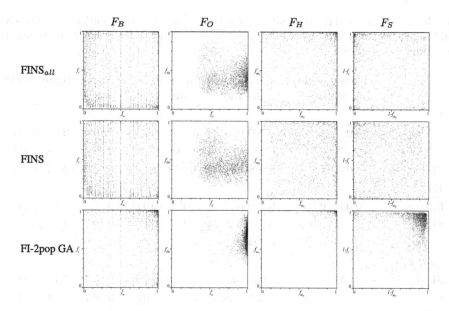

Fig. 4. Scatter plots of the visual metrics of all final spaceships in 50 independent runs, per evolutionary process and fitness function.

results for F_H are much more concentrated near $(1,1)$ than for F_S, while for F_O most spaceships have high scores in f_o but not particularly high scores in f_{bb}. To a degree, this is due to the fact that no sprite component fully covers its own bounding box, but also due to the fact that "boxy" spaceships of high f_{bb} tend to have simple outlines (f_o); the conflicting nature of these metrics allows f_o to dominate evolution as it is easier to optimize.

For FINS and FINS_{all}, the space of the two visual dimensions is more broadly explored, with interesting differences in the spread of results depending on both the approach and the visual dimensions. For instance, while for F_B a very broad range of f_t and f_w values are discovered, for F_O only average scores in both f_o and f_{bb} are discovered. Both novelty search approaches tend to concentrate on specific areas of the search space while exploring others to a smaller extent: for F_B more spaceships have low scores in f_t and f_w than high scores (while many spaceships also have $f_w = 0.5$, i.e. spaceships with square bounding boxes), for F_S more spaceships have low scores for $1 - f_{m_y}$, and (unsurprisingly) high scores in f_{m_y} and/or f_{m_x} for F_H. The bias towards certain areas of the search space is largely due to the representation and mutation operators used: due to the initial centrally placed sprite component and mutation operators which expand it vertically and horizontally, spaceships often even serendipitously receive high f_{m_y} and f_{m_x} scores, which is the case for both F_S and F_H. A visual inspection of Fig. 4 suggests that the targeted FINS explores more of the space of those dimensions included in its visual distance function than FINS_{all}, which explores more visual properties not included in the plot; this observation is tested numerically next.

Table 2. Diversity and cluster quality of the final spaceships per evolutionary approach and fitness function. Results are averaged from 50 independent runs with standard deviation in parentheses.

	Nearest neighbor distance	Inter-cluster distance	Intra-cluster distance	Cluster size	Dunn index
F_B					
FINS$_{all}$	0.065 (0.009)	0.700 (0.078)	0.124 (0.018)	6.9 (0.6)	2.26 (0.65)
FINS	0.068 (0.011)	0.639 (0.084)	0.135 (0.013)	7.4 (0.7)	2.01 (0.57)
FI-2pop GA	0.041 (0.013)	0.626 (0.108)	0.07 (0.025)	7.4 (0.7)	3.24 (1.09)
F_O					
FINS$_{all}$	0.037 (0.007)	0.329 (0.062)	0.059 (0.012)	6.9 (0.6)	2.13 (0.74)
FINS	0.043 (0.006)	0.331 (0.050)	0.071 (0.009)	7.2 (0.7)	1.88 (0.48)
FI-2pop GA	0.020 (0.007)	0.224 (0.080)	0.028 (0.009)	7.1 (0.9)	2.96 (1.47)
F_H					
FINS$_{all}$	0.049 (0.011)	0.655 (0.092)	0.088 (0.015)	6.9 (0.6)	2.78 (0.88)
FINS	0.063 (0.011)	0.594 (0.091)	0.117 (0.015)	6.9 (0.7)	2.11 (0.71)
FI-2pop GA	0.017 (0.007)	0.299 (0.115)	0.022 (0.013)	7.8 (0.5)	4.3 (4.55)
F_S					
FINS$_{all}$	0.056 (0.013)	0.681 (0.078)	0.104 (0.02)	6.9 (0.6)	2.44 (0.88)
FINS	0.069 (0.010)	0.606 (0.079)	0.127 (0.014)	7.2 (0.7)	1.87 (0.48)
FI-2pop GA	0.044 (0.015)	0.548 (0.129)	0.064 (0.024)	7.3 (0.7)	3.39 (1.27)

In order to evaluate the diversity of results more methodically, each of the 50 independent runs is analyzed individually: the nearest neighbor distance of all spaceships of the same run is averaged and included in Table 2. Moreover, the final spaceships of the same run are clustered via k-medoids to produce six clusters[1]. The average inter-cluster distance (i.e. the mean distance between all clusters' medoids), the average intra-cluster distance (i.e. the mean distance between all members of the same cluster) and the average size of the cluster are used to evaluate the clusters' quality. Finally, the clusters are evaluated via the Dunn index [27], which identifies how compact each cluster is and how separated different clusters are (by dividing the shortest inter-cluster distance by the longest intra-cluster distance). All of these metrics are included in Table 2. When calculating distances and for clustering, only the first two visual dimensions of the style objective functions of Eqs. (3)–(6) are considered; f_c is omitted as it does not have much visual impact and as its value range is similar in all experiments.

Unsurprisingly, the FI-2pop GA has a significantly lower nearest neighbor distance than both FINS and FINS$_{all}$, which is obvious from Fig. 4. For similar

---

[1] Due to stochasticity, the chosen clusters are the best of 100 independent clustering attempts.

reasons, the cluster medoids of spaceships evolved via FI-2pop GA are significantly closer together than those in FINS and FINS$_{all}$ (except in F_B). Surprisingly, the cluster quality based on the Dunn index is significantly higher for FI-2pop GA than for either novelty search approach; while the inter-cluster distance is relatively low, the intra-cluster distance is even lower (as all individuals are found in the same area of the search space, especially for F_H). Despite the fact that the clusters in the FI-2pop GA are not very separated, they are compact. Finally, regarding the clusters' size, all three evolutionary approaches had similarly sized clusters: many runs resulted in single-member clusters, as well as large ones. Clusters for the FI-2pop GA had more extreme size deviations, likely due to lower-quality outliers (caused by mutation) creating single-member clusters.

Comparisons between FINS and FINS$_{all}$ yield more interesting results. The dispersed appearance of FINS plots in Fig. 4 is validated via the nearest neighbor distance metric, which is significantly higher than FINS$_{all}$ in all cases except F_B. However, the k-medoid clusters for FINS$_{all}$ have a significantly higher inter-cluster distance than FINS in all cases except F_O. Individual spaceships are further dispersed with FINS (which rewards diversity along the same axes being evaluated), as evidenced by a high nearest neighbor distance and high intra-cluster distance, but the broader exploration of FINS$_{all}$ via a more granular distance function results in more easily separable and concise clusters, as evidenced by a high inter-cluster distance and a high Dunn index.

Figure 5 shows the cluster medoids of the run with the highest inter-cluster distance. The cluster medoids for the FI-2pop GA appear more similar; most medoids exhibit the intended visual style specified in the style objective, although major exceptions are the F_B objective which includes spaceships with wide, narrow and square bounding boxes as well as F_O which includes spaceships

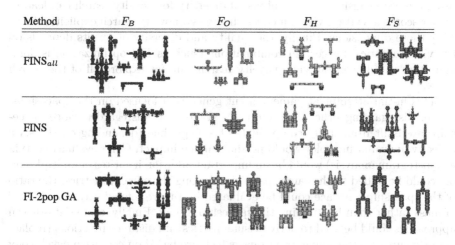

Method	F_B	F_O	F_H	F_S
FINS$_{all}$				
FINS				
FI-2pop GA				

Fig. 5. Cluster medoids of the independent runs with the highest inter-cluster distance (out of 50).

which do not cover much of the bounding box but exhibit complex outlines. Even so, cluster medoids of the FI-2pop GA exhibit at least one (and often both) visual properties in the objective, pointing to an overall consistent visual style in their results. On the other hand, medoids of either FINS or FINS$_{all}$ are visually very distinct (with few exceptions); it is important to note that even so, the medoids are different from one style objective to the next as well, confirming the impact that the visual metrics have on clustering. Taking F_S as an example, some medoids of FINS and FINS$_{all}$ have large side-sponsons while others are triangular, centrally aligned (horizontally and vertically), or front-loaded. Some medoids are more similar to each other, while others are unique. Overall, the cluster medoids for the different approaches are an indication of the range of visual styles which can be achieved via constrained novelty search, as well as the visual consistency (in most cases) which can be accomplished with the FI-2pop GA.

5   Discussion

Experiments in this paper evaluated the performance of two-population constrained evolutionary methods at generating game content of a designer-specified visual style as well as content of different, distinct visual styles. The objective-driven FI-2pop GA is ideal for creating spaceships of a visual style specified as an objective function, and the vast majority of its generated results possess the desired visual properties. However, in many cases the single-minded search of the FI-2pop GA can create "bizarre" spaceships which over-optimize the objectives (leading e.g. to the extremely wide spaceships for F_B in Fig. 3). Novelty search applied to feasible individuals creates visually different spaceships, yet there is no guarantee that most of those will score highly in a style objective. Surprisingly, using a more inclusive measure of visual diversity for novelty search can lead to higher scores in certain visual metrics than when novelty search explicitly diversifies those dimensions. Finally, the quality and diversity of results depends on the visual metrics: for certain combinations (such as F_O) optimization is dominated by one dimension while novelty search can not explore all of the search space.

This initial attempt at game content generation focused on the spaceships' shape, disregarding other characteristics. Color was disregarded since spaceships are built from sprite components of a single hue (depending on the style objective); evaluating color would not have much impact. Color evaluations (via e.g. histogram analysis) would be an important addition if sprites of multiple colors could be added to the same spaceship. Among the current metrics, the ratio of the spaceship's surface could be evaluated in a more refined way via a center of mass rather than by dividing the spaceship in half. Finally, more data-driven approaches could be used to derive visual metrics and distance functions (replacing or complementing the current ones which are based on cognitive psychology theory), such as object detection libraries for computer vision, e.g. in [28], or deep learning methods trained on previously generated content, e.g. in [29].

Results in this paper show that the generator's expressivity and the visual range of resulting content is sensitive to the design of both the visual metrics and the mutation operators. As noted in Sect. 3.2, the mutation operators were carefully crafted to avert infeasible spaceships, but also bias generation towards specific styles such as the centrally aligned spaceships of F_H. This affects novelty search which does not explore the search space as broadly as intended. Additionally, most visual metrics needed to be normalized via Eq. (2), as initial experiments with aggregated metrics in a style objective function indicated that some metrics tended to dominate others. These design choices, coupled with certain ad-hoc decisions (e.g. splitting the spaceship in halves) could affect the conclusions drawn from this study. Future work should improve and expand the mutation operators and methodically test the visual metrics.

While this paper focused almost exclusively on the visual style of the generated content, driven in part by a lack in the PCG literature in visual quality assessment, the functional properties of these spaceships should also be considered in future work. Since the generated spaceships could be used as enemies for an arcade-style game such as *Raptor: Call of the Shadows* (Apogee 1994), functional concerns could include how well the spaceships can fly or how dangerous they are. Currently, functional properties are considered only in constraints c_t and c_w, which ensure that the spaceship can fly and shoot; however, additional constraints on a minimum number of weapon components or a minimum and maximum speed for the spaceship would ensure that spaceships without thrusters (as in instances of Fig. 3) do not find their way into a game. For feasible spaceships, additional evaluations of their in-game performance (and game balance) would require simulations in a game engine, and could maximize their performance in pre-determined tasks (e.g. flying, avoiding obstacles, chasing enemies), similar to [22].

6 Conclusion

This paper described a method for evolving game content via turtle graphics where human-authored sprites are combined to create a complete, computer-generated spaceship. Generated results are evaluated on plausibility constraints and on visual metrics of the spaceship shape's balance and complexity. Two constrained evolutionary approaches, a feasible-infeasible two population genetic algorithm and feasible-infeasible novelty search, were tested on whether they generate game content of a specified visual style as well as content of different, distinct visual styles. Results indicate that while objective-driven search can create spaceships with a consistent visual style, the enhanced exploration of visual metrics afforded by novelty search can lead to interesting content which do not suffer from over-optimizing the engineered metrics of visual quality. Future work could improve the visual metrics, potentially replacing them with data-driven models, and include evaluations of the playability of generated content.

Acknowledgements. The sprite components used to construct the spaceships are freely licensed art assets found in OpenGameArt (http://opengameart.org/content/modular-ships) and are not the intellectual property of the author. The research was supported, in part, by the FP7 Marie Curie CIG project AutoGameDesign (project no: 630665).

References

1. Liapis, A., Yannakakis, G.N., Togelius, J.: Computational game creativity. In: Proceedings of the Fifth International Conference on Computational Creativity (2014)
2. Champandard, A.: EA investing in the next-generation with proceduralism post-SSX. http://aigamedev.com/open/teaser/investing-in-procedural/. Accessed 11 January 2016
3. Entertainment, B.: What is Diablo III? http://us.battle.net/d3/en/game/what-is. Accessed 11 January 2016
4. Togelius, J., Nelson, M.J., Liapis, A.: Characteristics of generatable games. In: Proceedings of the FDG Workshop on Procedural Content Generation (2014)
5. Arnheim, R.: Art and Visual Perception: A Psychology of the Creative Eye, Revised and expanded edn. University of California Press, Berkeley (2004)
6. Kimbrough, S.O., Koehler, G.J., Lu, M., Wood, D.H.: On a feasible-infeasible two-population (FI-2Pop) genetic algorithm for constrained optimization: Distance tracing and no free lunch. Eur. J. Oper. Res. **190**(2), 310–327 (2008)
7. Liapis, A., Yannakakis, G.N., Togelius, J.: Enhancements to constrained novelty search: Two-population novelty search for generating game content. In: Proceedings of Genetic and Evolutionary Computation Conference (2013)
8. Lehman, J., Stanley, K.O.: Revising the evolutionary computation abstraction: Minimal criteria novelty search. In: Proceedings of the Genetic and Evolutionary Computation Conference (2010)
9. Preuss, M., Liapis, A., Togelius, J.: Searching for good and diverse game levels. In: Proceedings of the IEEE Conference on Computational Intelligence and Games (CIG) (2014)
10. Smith, G., Whitehead, J.: Analyzing the expressive range of a level generator. In: Proceedings of the FDG workshop on Procedural Content Generation (2010)
11. Togelius, J., Yannakakis, G.N., Stanley, K.O., Browne, C.: Search-based procedural content generation: a taxonomy and survey. IEEE Trans. Comput. Intell. AI Games **3**(3), 172–186 (2011)
12. Liapis, A., Yannakakis, G.N., Togelius, J.: Towards a generic method of evaluating game levels. In: Proceedings of the AAAI Artificial Intelligence for Interactive Digital Entertainment Conference (2013)
13. Lara-Cabrera, R., Cotta, C., Fernández-Leiva, A.J.: A procedural balanced map generator with self-adaptive complexity for the real-time strategy game planet wars. In: Esparcia-Alcázar, A.I. (ed.) EvoApplications 2013. LNCS, vol. 7835, pp. 274–283. Springer, Heidelberg (2013)
14. Sorenson, N., Pasquier, P., DiPaola, S.: A generic approach to challenge modeling for the procedural creation of video game levels. IEEE Trans. Comput. Intell. AI Games **3**(3), 229–244 (2011)
15. Liapis, A., Yannakakis, G.N., Togelius, J.: Sentient Sketchbook: Computer-aided game level authoring. In: Proceedings of the 8th Conference on the Foundations of Digital Games (2013)

16. Howlett, A., Colton, S., Browne, C.: Evolving pixel shaders for the prototype video game subversion. In: Proceedings of the AI and Games Symposium (AISB 2010) (2010)

17. Hastings, E.J., Guha, R.K., Stanley, K.O.: Automatic content generation in the galactic arms race video game. IEEE Trans. Comput. Intell. AI Games 1(4), 245–263 (2009)

18. Risi, S., Lehman, J., D'Ambrosio, D., Hall, R., Stanley, K.O.: Combining search-based procedural content generation and social gaming in the petalz video game. In: Proceedings of Artificial Intelligence and Interactive Digital Entertainment Conference (2012)

19. Liapis, A., Yannakakis, G.N., Togelius, J.: Adapting models of visual aesthetics for personalized content creation. IEEE Trans. Comput. Intell. AI Games 4(3), 213–228 (2012)

20. Michalewicz, Z.: Do not kill unfeasible individuals. In: Proceedings of the Intelligent Information Systems Workshop (1995)

21. Coello Coello, C.A.: Constraint-handling techniques used with evolutionary algorithms. In: Proceedings of the Genetic and Evolutionary Computation Conference, ACM (2010)

22. Liapis, A., Yannakakis, G.N., Togelius, J.: Neuroevolutionary constrained optimization for content creation. In: Proceedings of the IEEE Conference on Computational Intelligence and Games (2011)

23. Lehman, J., Stanley, K.O.: Abandoning objectives: Evolution through the search for novelty alone. Evol. Comput. 19(2), 189–223 (2011)

24. Liapis, A., Yannakakis, G.N., Togelius, J.: Constrained novelty search: a study on game content generation. Evol. Comput. 23(1), 101–129 (2015)

25. Dunn, O.: Multiple comparisons among means. J. Am. Stat. Assoc. 56, 52–64 (2012)

26. Shapiro, S.S., Wilk, M.B.: Analysis of variance test for normality (complete samples). Biometrika (1965)

27. Dunn, J.C.: A fuzzy relative of the ISODATA process and its use in detecting compact well-separated clusters. J. Cybern. 3(3), 32–57 (1973)

28. Correia, J., Machado, P., Romero, J., Carballal, A.: Evolving figurative images using expression-based evolutionary art. In: Proceedings of the International Conference on Computational Creativity (2013)

29. Liapis, A., Martínez, H.P., Togelius, J., Yannakakis, G.N.: Transforming exploratory creativity with DeLeNoX. In: Proceedings of the International Conference on Computational Creativity (2013)

Grammatical Music Composition
with Dissimilarity Driven Hill Climbing

Róisín Loughran$^{(\boxtimes)}$, James McDermott, and Michael O'Neill

Natural Computing Research and Applications Group, University College Dublin,
Dublin, Ireland
roisin.loughran@ucd.ie

Abstract. An algorithmic compositional system that uses hill climbing
to create short melodies is presented. A context free grammar maps each
section of the resultant individual to a musical segment resulting in a
series of MIDI notes described by pitch and duration. The dissimilarity
between each pair of segments is measured using a metric based on the
pitch contour of the segments. Using a GUI, the user decides how many
segments to include and how they are to be distanced from each other.
The system performs a hill-climbing search using several mutation operators to create a population of segments the desired distances from each
other. A number of melodies composed by the system are presented that
demonstrate the algorithm's ability to match the desired targets and the
versatility created by the inclusion of the designed grammar.

Keywords: Algorithmic composition · Hill-climbing · Grammar

1 Introduction

This study introduces an algorithmic compositional system that creates short
melodies from the combination of a number of melodic phrases or segments
using a pre-defined context-free grammar. Each segment is related to each other
according to a distance specified by the user. The user need not define any
musical criteria for the compositions, merely how many segments there should
be and how similar each segment should be to each other. In this way, the user
can create compositions with no prior musical knowledge.

The proposed system is based on an evolutionary strategy. It is not dependent
on the 'survival of the fittest' concept of traditional evolutionary algorithms as
it does not measure fitness from individuals in a population but rather from the
combination of segments of one single individual. Thus a segment has no merit
on its own but only has importance according to how it is placed in relation to
its neighbours. Once an operation is performed, a newly created segment only
survives to the next generation if its inclusion improves the performance of the
whole individual. This improvement is measured in relation to how the segments
of the individual conform to the user-specified distance. The system employs a
hill climbing evolutionary strategy with variable neighbourhood search.

© Springer International Publishing Switzerland 2016
C. Johnson et al. (Eds.): EvoMUSART 2016, LNCS 9596, pp. 110–125, 2016.
DOI: 10.1007/978-3-319-31008-4_8

The following section describes some relevant previous work in the fields of algorithmic composition and background to this work. Section 3 describes the workings of the system, specifically the Grammar, Fitness Function and Operators used. A description of the experiments undertaken is given in Sect. 4. The results are discussed in Sect. 5 along with a selection of compositions before conclusions and future work are proposed in Sect. 6.

2   Related Work

Algorithmic composition involves composing music according to a given set of instructions or rules. While this could refer to any hand-written set of rules used to compose, in recent years it has involved the use of computer code and in particular machine learning techniques. A comprehensive survey of computational and AI techniques applied to algorithmic composition is given in [6].

2.1   Composing with Evolutionary Computation

This system is based on an evolutionary strategy. Evolutionary techniques are well suited to creative tasks such as music composition as they are population based and inherently non-deterministic; a solution is not determined outright but found by stochastically combining and altering high-performing solutions. Various EC methods have been used previously for melodic composition. Gen-Jam is a well-known system that uses a Genetic Algorithm (GA) to evolve jazz solos and has been used in live performances in mainstream venues [1]. A modified GA was used in GeNotator [19] to manipulate a musical composition using a hierarchical grammar. More recently, GAs were used to create four part harmonies without user-interaction or initial material according to rules from music theory [5]. Genetic Programming (GP) has been used to recursively described binary trees as genetic representation for the evolution of musical scores. The recursive mechanism of this representation allowed the generation of expressive performances and gestures along with musical notation [4]. Grammars were used with Grammatical Evolution (GE) [2] for composing short melodies in Elevated Pitch [16]. From four experimental setups of varying fitness functions and grammars they determined that users preferred melodies created with a structured grammar. GE was again employed for musical composition using the Wii remote for a generative, virtual system entitled Jive [17]. This system interactively modified a combination of sequences to create melodic pieces of musical interest.

The proposed system is a development of earlier experiments that used GE to create short novel music pieces. These experiments employed a grammar to create individuals consisting of notes, turns, chords and arpeggios and measured the fitness of each individual according to a statistical measure of the resultant tonality or the Zipf's distribution of a number of musical qualities [11,12]. An interesting observation from these studies is that when similar (but not identical) individuals were concatenated together, certain themes or motifs emerged giving a new musicality to the composition. This was possible to exploit using GE as such evolutionary methods use a population of solutions, resulting in a number of highly fit individuals at the end of a run. For the current system, it was decided to exploit this relationship further, taking a measure of similarity between the musically mapped individuals as a fitness function for the entire population. This has been developed into a hill-climbing system with the individuals in GE replaced by segments and the population replaced by one complete individual.

2.2 Musical Fitness

If an EC system does not use an interactive fitness function, one of the most difficult aspects of using the system to compose music is in designing the fitness function; how can we attribute numerical merit to one melody over another? The given system exploits the aesthetic aspect of repetition within music. Repetition in music has been shown to have a profound affect on the enjoyment of music, both in the repetition of full pieces [7,13] and in the analysis of form and meaning within music [15]. Repetition was also found to be an extremely useful aspect of using EC in previous experiments [12] leading to a focus on this quality for the given system.

For these compositions it is the variation on a theme — segments of a piece that are partly repeated or share similarities in some respect — rather than an exact repetition of a phrase that is desired. To develop such a fitness function a method of measuring the distance between two given melodies is required. Many studies have looked at musical similarities, most concentrating on similarities in the audio signal [10,18]. Others, such as the current system, only consider the MIDI values of the notes. Some studies have used edit distances for such measurements. A probabilistic method on melodic similarity was used with a designed edit distance for query by humming tasks [8]. Measures of similarities between musical contours have also been considered [3].

The musical distance used in this system is measured from a step-wise list of the contour of the pitches in the segments as discussed in the following section.

3   Method

This section describes these three processes used in this system to compose music: the grammar, fitness measure and operators.

3.1 Grammar

Context free grammars can be used to map information from one domain into a more meaningful domain, allowing the user to develop a representation suitable for the problem at hand. Such a grammar is employed in these experiments similarly to how they are used in evolutionary methods such as GE, whereby the genotype — a variable length integer string known as codons, is mapped to the phenotype — a command language that is interpreted into MIDI notes. A grammar in Backus-Naur Form (BNF) is used, whereby a given expression is expanded according to a series of production rules. The choice of each rule is determined by the current integer codon:

$$\text{Rule} = (\text{Codon Integer Value})\,mod(\#\text{ of choices}) \qquad (1)$$

The creative capabilities of grammar-based methods come from the choices offered within the mapping of the grammar. The grammar used in this system expands the genotype into the description of a number of musical events — notes, chord, turns and arpeggios. The grammar and a brief description of its use is shown below. A more detailed description of the inner workings of this grammar can be found in [12].

```
<individual> ::= <piece>|<piece>|<piece><transpose>
<piece> ::= <note><note><note><note><note>
<note>::= 111,<style>,<oct>,<pitch>,<dur>,

<style>::= 100|100|100|100|100|100|100|100|50,<chord>|50,<chord>|
          50,<chord>|50,<chord>|70,<turn>,100|80,<arp>,100
<transpose> ::= 90,<dir>,<TrStep>,
<TrStep> ::= 0|1|2|2|3|3|4|5|5|5|6|7|7|7|8|8|9|9|10|10|11
<chord>::= <int>,0,0|<int>,<int>,0|12,0,0|<int>,0,0|<int>,0,0
          |<int>,0,0|<int>,<int>,<int>
<turn>::= <dir>,<len>,<dir>,<len>,<stepD>
<arp>::= <dir>,<int>,<dir>,<int>,<ArpDur>

<int>::= 3|4|5|7|5|5|7|7
<len>::= <step>|<step>,<step> |<step>,<step>,<step>
        |<step>,<step>,<step>,<step>|<step>,<step>,<step>
<dir>::= 45|55
<step>::= 1|1|1|1|1|1|2|2|2|2|2|2|2|2|3
<stepD>::= 1|2|2|2|2|2|2|4|4|4|4|4|4
<ArpDur>::= 2|2|2|4|4|4|4|4|8|8
<oct>::= 3|4|4|4|4|5|5|5|5|6|6
<pitch>::= 0|1|2|3|4|5|6|7|8|9|10|11
<dur>::= 1|1|1|2|2|2|4|4|4|8|8|16|16|32
```

This grammar results in an individual piece of music that may be transposed up or down a given interval. The piece is comprised of five note events each of which can either be a single note, a chord, a turn or an arpeggio. A single note is described by a given pitch, duration and octave value. A chord is given these values but also either one, two or three notes played above the given note at specified intervals. Turn results in a series of notes proceeding in the direction up or down or a combination of both. Each step in a turn is limited to either one, two or three semitones. An arpeggio is similar to a turn except it allows larger intervals and longer durations. The application of this grammar results in a series of notes each with a given pitch and duration. The inclusion of turns and arpeggios allows a variation in the number of notes that are played, depending on the production rules chosen by the grammar. The use of such grammars allows the introduction of a bias by including more instances of one choice over another. For example, a tone (value 2) is the most likely choice for the <step> rule above as there are more instances of that choice available.

3.2   Fitness Measurement

The fitness of the individual is calculated after the grammar has been employed to expand each segment into a series of notes. Fitness is taken as a measure of how close the segments fit to a pre-chosen pattern within a metric space.

Fig. 1. Four variations on a melody

To consider the distance between two segments, the relationship between the pair is examined at every time-step. The pitch line is expanded for each segment to give a pitch value at each demisemiquaver (duration of 1 in the grammar). For example, if there is a crotchet (duration 8) played at D (pitch 2) this is represented as a list of 8 values of 2. This results in a list for each segment that indicates the pitch of that segment at every moment of duration. In the case of a chord, only the root note of the chord is considered. These experiments encourage the use of transpositions of melodies; melodies that are alike apart from a pitch shift are to be deemed equal. To achieve this, each pitch vector is normalised to start at 0. In this manner it is a *pitch contour* that is being examined, rather than the actual pitch values. The distance between two segments is then taken as the sum of the absolute distance between their pitch contours at each time step. If one contour is longer than the other, the difference in length between the two is multiplied by 5 and added to the distance between them. As there is a maximum distance of 11 semitones between two lines at any point, the value of 5 was chosen as the median of this pitch interval. This heavily penalises segments of different lengths to account for possible pitch differences as well as the length difference when notes are absent. To illustrate this, consider the four musical measures depicted in Fig. 1. As described, these contours are calculated as:

ContourA [0 0 0 0 4 4 6 6 7 7 7 7 6 6 6 6]
ContourB [0 0 0 0 4 4 6 6 7 7 7 7 5 5 5 5]
ContourC [0 0 0 0 0 0 0 0 7 7 7 7 6 6 6 6]
ContourD [0 0 0 0 4 4 4 4 6 6 6 6 7 7 7 7 6 6 6 6]

The differences between each variation to the original melody A is given by:

Diff A−B sum[0 0 0 0 0 0 0 0 0 0 0 0 1 1 1 1] = 4
Diff A−C sum[0 0 0 0 4 4 6 6 0 0 0 0 0 0 0 0] = 20
Diff A−D sum[0 0 0 0 0 0 2 2 1 1 1 1 1 1 1 1] + (5 × 4) = 32

From these measurements, Melody B is most similar to Melody A, followed by C and finally D. Using this distance metric it can be ensured that melodies which stray in pitch but maintain a similar rhythm (such as Melody A & B) are not considered as different as those that differ in rhythm (such as Melody A & C). Two melodies which are identical other than an early timing difference will be measured as having a large dissimilarity between them: a 'ripple effect' of the timing difference will cause all following notes in one melody to be offset from their matching notes in the other. We acknowledge that this is only one possible measure of melodic similarity and are considering alternative methods. The use of dynamic programming algorithms such as the Levenshtein edit distance [9] were also considered for this purpose, although early experiments did not produce encouraging results with this system.

The distance between each pair of segments in the individual is measured and compared to an ideal list of measurements to determine fitness. We acknowledge that the space within which the distances are measured is an abstract metric space. However, it is useful for the user to visualise these distances in a two-dimensional geometric space. For example, consider five segments [a, b, c, d, e] equally spaced in a pentagonal shape as per the diagram in Fig. 2. We know that in a pentagon the length of a diagonal e.g. (a, c) is approximately 1.6 times the length of a side e.g. (a, b). The spacing of the segments can be completely described by specifying a distance between each pair of points:

```
      a   b   c   d   e
a   0.0 1.0 1.6 1.6 1.0
b   1.0 0.0 1.0 1.6 1.6
c   1.6 1.0 0.0 1.0 1.6
d   1.6 1.6 1.0 0.0 1.0
e   1.0 1.6 1.6 1.0 0.0
```

As this is symmetrical about the diagonal, a unique description of the distances between all points can be given by the upper triangle of this matrix collapsed into the row vector. Hence the pentagon shown can be described by the distances [1, 1.6, 1.6, 1, 1, 1.6, 1.6, 1, 1.6, 1] or a scalar multiple of this vector. The fitness of an individual is measured by defining a list of ideal distances such as this and taking the absolute error from the actual measured distances to these ideal distances for each pair of segments. The fitness of the individual is calculated as the mean squared error (mse) of this list of errors.

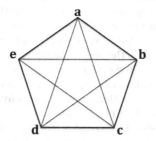

Fig. 2. Five segments equally spaced in a pentagonal arrangement

Each individual is defined as an ordered list of segments. Due to the method of calculating fitness, the ordering of these segments is extremely important. Re-ordering the given segments is one of the simplest yet most powerful operations that can be implemented on the individual melody, as described in the following section.

3.3   Operators

In each generation there are five possible operators performed on the individual: Switch, Permute, Crossover, Mutate and Copy. Each of these are performed in succession until an improvement on the individual is found. Once the fitness improves, the individual is updated and the next generation is started.

Switch. The Switch operator switches the ordering of each pair of segments. A new individual fitness is measured for each switched pair and the best fitness obtained is recorded.

Permute. The Permute operator makes one random permutation of all of the ordered segments. The ordering and fitness of the new individual is noted.

Crossover. The Crossover operator selects each pair of segments within the individual and performs typical crossover on them, resulting in a new pair of segments. Again the new individual fitness is measured and recorded for each crossed pair and the best fitness is noted.

Mutate. The Mutate operator selects each segment in turn and performs a mutation on it. The mutation rate is initially set to 0.01 giving a 1 % chance that any given codon will be mutated. The new individual and fitness is measured and recorded for each mutation.

Copy. The Copy operator picks two segments at random and replaces one with a copy of the other. The new individual and fitness are noted.

Crossover and Mutate are typical operators in any evolutionary system. Switch and Permute were developed for this particular system as the ordering of the segments has been shown to be very important. Copy was included as repetition or variation of a segment may be very important in the system. These operators are implemented in order of how much effect they have on the individual. Switch merely switches the order of two segments and so has least effect on the individual. Permute has more effect in that it can re-order all segments, but it still does not introduce any new material to the individual. Mutate and Crossover both introduce new material. Mutate replaces one segment, whereas Crossover combines two segments into two new segments. The degree to which each of these operators has an effect on the content of the individual is dependant on the mutation rate and the selection point for crossover. With a high mutation rate a single mutation can cause a large difference in a segment. Conversely if the crossover location is towards the end of the genome, much of the initial segments may remain unchanged. Hence Crossover is considered to have less effect on the individual, although this is dependent on the specific operation. Copy completely removes one segment from the individual without introducing any new material and is hence considered to have the largest effect on the individual.

With these operators applied in succession, the algorithm continually implements changes of a small step size until this small step is no longer effective. Only then can it make a larger change to the individual. Once any change is made, the algorithm returns to checking each Switch again for a small improvement. In this way, the algorithm systematically searches a small step — or the local search space — before searching further away for a better performance. This operates in the manner of a Variable Neighbourhood Search (VNS) whereby a local minimum is found before the search moves out to a wider neighbourhood [14]. The operators that perform multiple operations (Switch, Crossover, Mutate) systematically try each operational possibility and pick the best improvement as the outcome. Conversely Permute and Copy are only given one chance in each generation to improve the fitness. Permute is used this way as it is considered an extension of Switch. Similarly, Copy is only used once at the end of a generation when no other operator has improved the fitness.

4 Experiments

This section discusses the design choices considered in implementing the system.

4.1 Melody Shapes

For a user-friendly system, the user must be able to easily specify distances between segments. The simplest way to define such distances is graphically. A 'Music Geometry GUI' was created that allows the user to specify the number of segments and their placements in a two dimensional square. The user can place any number of points within the square and a numerical distance (integer) between each pair of points is returned. The target vector in the fitness function is this list of distances. The length of the composed melody is dependant on both the number and length of segments used. The number of segments in an individual melody is controlled by the user as the number of points she plots. The length of each segment can be controlled by adding more instances of <note> in the grammar. In either method of elongating the composition, more calculations are involved and hence a longer run time is required.

Four separate shapes were defined using the Music Geometry GUI. The design of these shapes is shown in Fig. 3. Cluster is the simplest shape containing only six segments split into two distinct clusters. Circle contains nine segments that traverse in a circular shape before returning to where they started. Cross also contains nine segments, whereby each alternative segment returns to the original starting point, forming a cross shape. Line contains twelve segments at similar distances moving linearly away from the starting point.

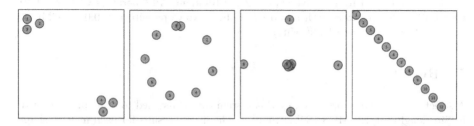

Fig. 3. Shape Targets for the Cluster, Circular, Cross and Line melodies as created by the Music Geometry GUI

4.2 Experimental Variations

Each experiment was run 30 times on all shapes with five set-ups.

Random Search. To confirm the search operators were effective on the given problem, each experiment was compared against a random search for the desired shape. For this, a new individual was initialised and evaluated a given number of times, recording the best fitness achieved. To ensure a fair comparison, the number of evaluations was adjusted according to the number of segments in each shape. For the Cluster melody, six segments results in 15 evaluations for both the Switch and Crossover operators, six evaluations for the Mutate operator and Permute and Copy will make one evaluation each resulting in up to 38 fitness evaluations per generation. Likewise for the Circle and Cross individuals, both with nine segments, there can be up to 83 evaluations $(36 + 36 + 9 + 1 + 1)$ and Line individuals could require up to 146 $(66 + 66 + 12 + 1 + 1)$. As each experiment was run for 1,000 generations, individuals were initialised and evaluated in the random search for Cluster, Circle, Cross and Line 38,000, 83,000, 83,000 and 146,000 times respectively.

Mutation Coefficient. The first experiment incorporates all operators with the Mutation Rate (μ) set to 0.01. This is a typical μ value in GE experiments employing a BNF grammar with an initial number of 100 codons. A higher μ typically results in random-like behaviour in GE experiments due to the ripple effect in which changes in the genome can have a very destructive effect on the phenome. Nevertheless, in preliminary studies it was found that using a higher μ in these experiments was leading to better final fitness. Mutation operates in a different manner in this system as regardless of the value of μ, only one segment is mutated, hence only part of the individual is affected. As a second experiment μ was increased to 0.1 and melodies for each shape were evolved again.

Limiting Operators. The final operator used in any run is Copy. This operator removes one entire segment without introducing new material so it could be very destructive. Conversely, for patterns that require two segments to be very close or identical, Copy may be very beneficial. To examine the effectiveness of Copy within the system each experiment was run again without the use of Copy.

This results in a total of five experiments: Random, All Operators with $\mu = 0.01$ (AllMu01), All Operators with $\mu = 0.1$ (AllMu1), No Copy with $\mu = 0.01$ (NCMu01) and No Copy with $\mu = 0.1$ (NCMu1).

5 Results

A selection of melodies produced by this system can be listened to at http://ncra.ucd.ie/Site/loughranr/EvoMUSART_2016.html. The fitness results for each melody shape both for all four variations of the system and for random search are shown in Fig. 4. The fitness results shown are the average of the best fitnesses achieved over 30 runs. These figures demonstrate a consistent performance for each experimental variation on each melody pattern. It is clear that each version of the system achieves better fitness than random search for each melody pattern. It is also evident that the AllMu01 and

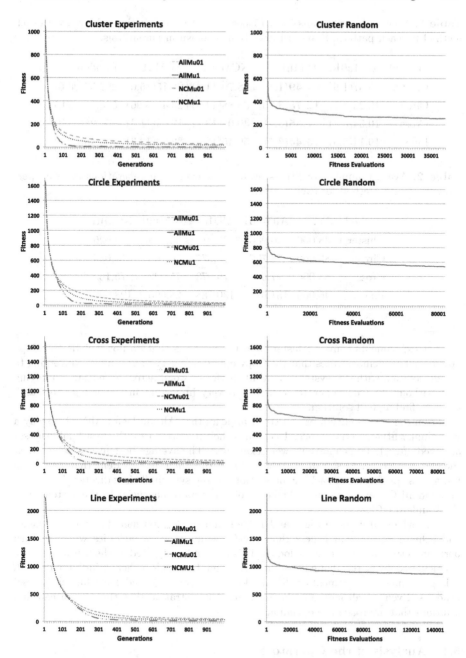

Fig. 4. Average best fitness for each experimental run and Random Initialisation for each melody shape. Note scales between shapes will differ due to different number of segments in each shape (Color figure online).

Table 1. Average (over 30 runs) best Fitness after 1000 generation achieved with each method for each pattern. Standard deviation is shown in parenthesis.

Method	AllMu01	AllMu1	NCMu01	NCMu1	Random
Cluster	5.37(1.9)	**4.49**(1.8)	28.79(11.4)	21.15(15.0)	252.52(26.6)
Circle	13.23(4.5)	**12.78**(3.15)	35.88(14.5)	23.48(8.26)	535.861(54.7)
Cross	10.54(5.7)	**7.55**(3.4)	40.08(18.7)	17.89(11.3)	556.21(54.3)
Line	19.14(5.9)	**15.44**(3.7)	50.45(18.8)	32.73(13.3)	864.48(65.7)

Table 2. Average number of fitness evaluations over 1000 generations actually performed by each method for each pattern

Method	Max	AllMu01	AllMu1	NCMu01	NCMu1
Cluster	38,000	36,986	36,911	35,666	35,626
Circle	83,000	78,777	78,677	76,765	77,086
Cross	83,000	77,630	76,973	75,292	75,169
Line	146,000	131,618	130,788	128,695	128,998

AllMu1 experiments converge faster than the methods that do not include the Copy operator. This demonstrates that the Copy operator is very important in converging to a good solution with this system. NCMu01 performs worst across each shape, implying that without the Copy operator a high μ is very important in traversing the search space to find a good solution.

From the fitness curves presented, it appears that AllMu01 and AllMu1 display a very similar fitness performance. To confirm their performances, the best final fitnesses achieved after 1,000 generations were examined. The average best fitness achieved at the end of each run is shown in Table 1. From this table it is evident that AllMu1 is the highest performer of each of the versions of the system. This indicates that unlike in standard GE experiments, a higher μ of 0.1 is more beneficial to the system than the standard 0.01.

It is evident from both Fig. 4 and Table 1 that the Cluster melodies converge faster and achieve better final fitness than the Cross and Circles melodies which in turn perform better than the Line melodies. This is as to be expected as the fitness measure is calculated from the distance of each segment to an ideal, hence individuals with a larger number of segments will take longer to converge and have higher (worse) fitnesses. Even so, all methods do converge, demonstrating that the system can find solutions that approach the optimum.

5.1   Analysis of the Operators

As described in Sect. 3 the operators are applied according to a variable neighbourhood search leading to a difference in the number of fitness evaluations in each generation. Table 2 displays the average number of evaluations actually performed for each shape[1].

---

[1] Note that the NC methods use less as each Copy in each of the 1,000 generations requires one evaluation.

Table 3. Number of times each operator improved fitness with the AllMul experiment for each shape

Method	Switch	Permute	Cross	Mutate	Copy
Cluster	1122	18	664	687	280
Circle	1353	7	709	786	231
Cross	3500	2	1285	864	436
Line	5300	0	1987	1472	427

If after all operators have been performed there is no improvement, the current individual is kept until the next generation, resulting in the maximum number of fitness evaluations for that generation. The initial improvement in fitness shown during the evolution of all melodies demonstrates that the operators are successful and hence there are fewer fitness evaluations in early generations than in later ones. Certain operators were found to be more useful than others. The total number of times each operator produced a improvement in fitness is shown in Table 3.

It can be seen that for each shape Switch is the most successful operator. This is to be expected as it implements the smallest change on the individual and it is the first operator applied in each generation. Often Switch is applied many times in succession as when a good Switch is encountered, the individual is updated and then each Switch combination is applied again to see if further improvement can be found. This is seen to be particularly useful at the beginning of a run. Crossover and Mutate were found to be the next most successful. Copy was already shown to be beneficial in the experiments, however it is not found to be as successful as the previous operators. This is to be expected as it is quite destructive and also it is the last operator to be checked for a fitness improvement. Permute, however, is very rarely chosen as a successful operator. There are a number of possible explanations for this. Firstly, Permute is only given one random opportunity to produce a better individual whereas Switch, Cross and Mutate all have multiple opportunities depending on the number of segments in the individual. Furthermore, Permute is implemented directly after all possible Switch operators have been tested (i.e. the Switch arrangement is optimal). Hence a further re-ordering of the segments is unlikely to have an effect. Also it should be noted that Permute is used more for individuals with lower number of segments than with higher.

It was observed that more operators were successfully used at the beginning of a run. To test this observation the number of fitness evaluations made were compared against the highest number possible for the Line melodies. As each Line melody has 12 segments, the maximum number of fitness evaluations per generation is 146. Figure 5 shows this maximum against the number of evaluations actually implemented for each of the four methods on the Line melody. This figure shows that for each method the number of evaluations increases more slowly than the maximum up until approximately generation 300 at which point it increases in line with the maximum. As expected, this change in frequency of implementation of an operator aligns with the point of convergence of these runs, at which point little further improvement is seen in the fitness.

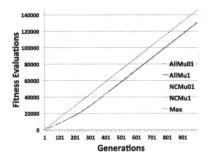

Fig. 5. Average number of fitness evaluations per generation for each of the four methods on the Line melody compared the maximum possible number

5.2   Melodies

From produced melodies, it is clear that some shapes are easier to hear than others. Of the four shapes, we found Cluster and Cross to be easiest to identify. These melodies are repeated twice to emphasise their corresponding shapes. The similarities in the segments within the clusters in the Cluster melodies are evident to the ear. Even when these are transposed, as in Cluster2, the transposed phrases are easy for the ear to recognise. Likewise with the Cross melodies, e.g. Cross1, the return to the central segment results in a cyclical aspect to the composition. The pattern in the Circle melodies are slightly harder to discern. Each Circle melody is repeated three times to encourage the pattern to emerge. Each segment steps away from the previous but circles around and returns to the starting position. As the piece gradually returns we can hear consistencies throughout the melody. Circle3 for instance is a slow moving piece with a semibreve in each segment. Although the segments go through many changes in the cycle, this long note anchors the piece into a sense of continuity. The combination of quick runs followed by a long chord results in a similar effect in Circle1. In the Line melodies the segments can be heard to move gradually away from where they started. However, without a return to a specific point such as with the Cross and Circle melodies, or a closer variation within a group such as with the Cluster melodies, the musical quality of these Line melodies is less evident than those produced from other shapes.

When the shape is known, the authors can generally 'hear' this shape. To determine if this is recognisable to a naive listener, a series of listening tests would need to be conducted. The purpose of the system is not to exactly match such patterns however, but to create an easy to use compositional system that can create interesting melodies with a specified or implied level of repetition. The grammar used and the measurement on a pitch contour lead to a transposed meaning of what constitutes the 'same' melody, but this is purposefully introduced into the system to maintain variety and diversity in the music produced.

5.3   Discussion

Evolutionary methods such as this hill-climbing algorithm are very suitable to this type of creative problem. The stochastic element inherent to EC combined with the specified grammatical mapping can result in a variety of musical features. It is difficult to formulate a linear or deterministic algorithm to create music that offers true surprise or novelty. EC offers a rich search space that can be traversed in a variety of ways depending on the fitness function selection methods used. This particular method uses the relationship between elements of the composition as a selection criteria rather than any specific musical content, thus the user can have no concrete expectations as to what content the system will produce. Such systems offer a balance between randomness and determinism in which computational creativity may have the opportunity to flourish.

As discussed, the above experiments were run over 1,000 generations resulting in a maximum of 38,000 to 146,000 fitness evaluations per run. This may be considered small in comparison to some EC experiments and yet the converged fitnesses in Fig. 4 show that this is sufficient. The lack of improvement in the Random plots however, prove that this is not a simple problem. This demonstrates that the system is efficient at finding good solutions and the operators chosen are adept at traversing the search space effectively.

The way in which the user controls the composition is somewhat abstract. The user has no control of the melodic content within the compositions. For the Line melodies the linear decrease in the shape does not imply a decrease in pitch, merely that the segments are linearly separated in the metric space. Likewise a circular pattern doesn't imply a circle of pitches. Nevertheless, we found that controlling this placement of segments can result in interesting compositions.

The purpose of the system is not to make musical shapes, as such shapes do not have particular meaning in a musical sense. The specific shapes described in this experiment are used to demonstrate the operation of the system. The system does however offer the user a simple way of controlling, albeit in an indirect manner, the approximate length of the compositions and how much repetition or similarity is contained within the melody. Combining the distance metric with the grammar results in a one-to-many mapping; the distance metric from one shape can create a wide variety of melodies. The system can be seen as either a potential compositional tool to assist in the creation of musical ideas or a fun way for people with no musical experience to create musical excerpts without having to specify any musical qualities in relation to timing or pitch.

6   Conclusion and Future Work

A system is presented that composes short melodies using a hill-climbing algorithm with a context-free grammar. The system is driven by a fitness function based on the distances between user-placed segments. No key or time signature is incorporated into the system and the user does not require any musical knowledge. A selection of melodies created by the system are presented demonstrating that the system can create a variety of melodies that correspond to the metric distances specified by the user. The success of the evolutionary runs of the system was confirmed against random search confirming that the defined operators were extremely adept at traversing the search space to find a good solution. By employing the operators in succession a variable neighbourhood search was adopted in which the Switch operator was found to be most successful in improving fitness followed by Crossover and Mutate.

In future work, we plan to consider alternative distance metrics for comparing segments and to consider larger-scale segments that can refer to form in larger compositions. We are planning a series of listening experiments on melodies created by the system to determine if there is a correlation between the similarity of melodic segments and musical preference. The music produced by the system is dependent on the designed grammar, the implemented operators, the similarity measure used in the fitness function and the shapes chosen by the user. In a future experiment we would like to isolate these various aspects to determine their individual effects of the produced melodies. In doing so, we hope to learn more about the perception and understanding of melodic expectation.

Acknowledgments. This work is part of the App'Ed (Applications of Evolutionary Design) project funded by Science Foundation Ireland under grant 13/IA/1850.

References

1. Biles, J.A.: Straight-ahead jazz with GenJam: a quick demonstration. In: MUME 2013 Workshop (2013)
2. Brabazon, A., O'Neill, M., McGarraghy, S.: Grammatical evolution. In: Brabazon, A., O'Neill, M., McGarraghy, S. (eds.) Natural Computing Algorithms, pp. 357–373. Springer, Heidelberg (2015)
3. Chen, A.L., Chang, M., Chen, J., Hsu, J.L., Hsu, C.H., Hua, S.: Query by music segments: an efficient approach for song retrieval. In: ICME 2000, vol. 2, pp. 873–876. IEEE (2000)
4. Dahlstedt, P.: Autonomous evolution of complete piano pieces and performances. In: Proceedings of Music AL Workshop. Citeseer (2007)
5. Donnelly, P., Sheppard, J.: Evolving four-part harmony using genetic algorithms. In: Di Chio, C., et al. (eds.) EvoApplications 2011, Part II. LNCS, vol. 6625, pp. 273–282. Springer, Heidelberg (2011)
6. Fernández, J.D., Vico, F.: AI methods in algorithmic composition: a comprehensive survey. J. Artif. Intell. Res. **48**, 513–582 (2013)
7. Hargreaves, D.J.: The effects of repetition on liking for music. J. Res. Music Educ. **32**(1), 35–47 (1984)
8. Hu, N., Dannenberg, R.B., Lewis, A.L.: A probabilistic model of melodic similarity. In: International Computer Music Conference (ICMC), Goteborg, Sweden. International Computer Music Society (2002)
9. Lemström, K., Ukkonen, E.: Including interval encoding into edit distance based music comparison and retrieval. In: Proceedings of the AISB, pp. 53–60 (2000)
10. Logan, B., Salomon, A.: A music similarity function based on signal analysis. In: ICME, Tokyo, Japan, p. 190. IEEE (2001)
11. Loughran, R., McDermott, J., O'Neill, M.: Grammatical evolution with Zipf's law based fitness for melodic composition. In: Sound and Music Computing, Maynooth (2015)
12. Loughran, R., McDermott, J., O'Neill, M.: Tonality driven piano compositions with grammatical evolution. In: CEC, pp. 2168–2175. IEEE (2015)
13. Middleton, R.: 'Play It Again Sam': some notes on the productivity of repetition in popular music. Popular Music **3**, 235–270 (1983)
14. Mladenović, N., Hansen, P.: Variable neighborhood search. Comput. Oper. Res. **24**(11), 1097–1100 (1997)
15. Ockelford, A.: Repetition in Music: Theoretical and Metatheoretical Perspectives. Ashgate, Aldershot (2005)
16. Reddin, J., McDermott, J., O'Neill, M.: Elevated pitch: automated grammatical evolution of short compositions. In: Giacobini, M., et al. (eds.) EvoWorkshops 2009. LNCS, vol. 5484, pp. 579–584. Springer, Heidelberg (2009)

17. Shao, J., McDermott, J., O'Neill, M., Brabazon, A.: Jive: a generative, interactive, virtual, evolutionary music system. In: Di Chio, C., et al. (eds.) EvoApplications 2010, Part II. LNCS, vol. 6025, pp. 341–350. Springer, Heidelberg (2010)
18. Slaney, M., Weinberger, K., White, W.: Learning a metric for music similarity. In: International Symposium on Music Information Retrieval (ISMIR) (2008)
19. Thywissen, K.: GeNotator: an environment for exploring the application of evolutionary techniques in computer-assisted composition. Organ. Sound 4(2), 127–133 (1999)

Animating Typescript Using Aesthetically Evolved Images

Ashley Mills[(✉)]

School of Computing, The University of Kent, Canterbury, UK
ashley@ashleymills.com, ajsm@kent.ac.uk

Abstract. The genotypic functions from apriori aesthetically evolved images are mutated progressively and their phenotypes sequenced temporally to produce animated versions. The animated versions are mapped onto typeface and combined spatially to produce animated typescript. The output is then discussed with reference to computer aided design and machine learning.

Keywords: Aesthetic evolution animated · Animated typeface · Typescript

1 Introduction

Systems for evolving aesthetic images, whilst still unfamiliar in popular culture, are academically well known and date back at least 20 years [1–5]. Ordinarily the output of an image-centric aesthetic evolution system is seen as the end product in itself; aesthetically pleasing images are the goal. The genotypes of these images however encode the labour of a prolonged and iteratively applied intelligent cognitive filtering process. This serves to apriori mark out these artifacts as cognitively interesting and artistically compelling.

It is prudent therefore to make use of these creatively valuable artifacts in producing further creative output. With this motivation, in this work, the output of an Aesthetic Evolution (AE) system based on [6] is used to produce animated typefaces. This work focuses on using AE output to animate *existing* typefaces, which sets it apart from work such as [7] which focuses on the construction of a system to evolve novel typefaces.

The rest of this document is outlined as follows, in Sect. 2 we briefly describe the AE system which generates the image inputs for typeface animation, in Sect. 3 we explain how the images, or rather their genotypic functions, can be modulated to produce animation, in Sect. 4 we describe how these animations can be mapped to individual glyphs, and in Sect. 5 we explain how the last can be combined into typescript. Finally, in Sect. 6 we discuss the artistic applications and potential machine learning applications that this work inspires.

© Springer International Publishing Switzerland 2016
C. Johnson et al. (Eds.): EvoMUSART 2016, LNCS 9596, pp. 126–134, 2016.
DOI: 10.1007/978-3-319-31008-4_9

2   Aesthetic Evolution Basis

The aesthetic evolution system used here employs a classical Cartesian coordinate model [1] as follows. Images are represented genotypically as a triplet of function trees; one tree for each color in the RGB colorspace. The function trees are initially grown stochastically from a pool of primitive function nodes such as sin, cos, etc. with arguments that are themselves either recursively function nodes, or terminal primitives such as numerical constants or external variables. In the set of terminal primitives, the variables x and y exist to represent the Cartesian coordinates of the image and at least one must occur in each generated function tree to produce a valid image.

To produce an image, the genotypic function triplet is evaluated for each pixel, i.e. for each x and y coordinate of the output image, where x and y are normalized, irrespective of image size, to fall within the range [0,1]. Thus, the leftmost pixel of an image assumes the value $x = 0$ and the rightmost the value $x = 1$. The return values from the evaluated triplet provide the RGB values for each pixel of the image respectively. A trivial example of such a function triplet is shown in Fig. 1, along with its image. Note that the triplet members are the three branches that enter the function "new_hsv".

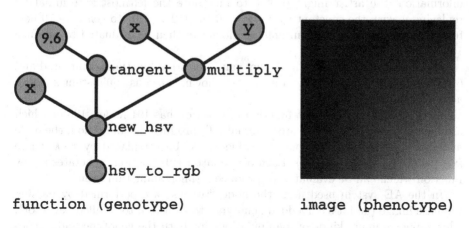

function (genotype)   image (phenotype)

Fig. 1. A simple functional triplet is shown on the left along with its phenotypic expression, the image, on the right. Note in this case, the triplet passes through an HSV colorspace before entering the RGB colorspace (Color figure online).

Successive generations of images are obtained in the well known manner of visual cortex guided manual selection, followed by randomly seeded mutation and crossover. After many such generations, the images produced converge on aesthetically pleasing forms. These images are used as the input to the animated typeface system described subsequently. For more information about the aesthetic evolution system used here, please see [6].

An interesting and useful feature of this AE system is that it outputs phenotype files whose genotypes are embedded within them: each image is output as a JPEG and the genotype is stored in a JPEG metadata segment (APP0 marker segment) of that image. By combining genotype and phenotype into a single file whose format conforms with an existing standard, the output images can be viewed and loaded by a plethora of existing software, which makes for convenient manipulation and organization. It also makes it a straightforward matter to obtain the genotype for an image when it is desired to manipulate the underlying form. To avoid namespace confusion, the format uses the extension "exp.jpg" meaning expression embedded JPEG instead of the usual ".jpg".

3   Animating Aesthetically Evolved Images

To animate a previously aesthetically evolved image, the underlying function needs be modified through time. This process is relatively straightforward and proceeds as follows. First, an image with appropriate artistic merit is selected and it's nodes are enumerated and examined. For example, the image in Fig. 1 has four function nodes "hsv_to_rgb", "new_hsv", "tangent", and "multiply" as well as four terminal nodes "x", "9.6", "x", and "y". After examining this information, the artist might decide to modulate the leftmost x terminal by replacing it with the constant values $x = \{0, 0.1, 0.2, ..., 1.0\}$ to generate 10 new functions. Each of the 10 resultant functions can then be evaluated to produce 10 animation frames (Fig. 2).

As an example consider Fig. 3 where an interesting function is animated and 6 frames from the animation sequence are shown, taken as equidistant samples across the animated range.

In this particular case the function in question has 191 nodes, 120 of which are functions and 71 of which are terminals. To produce the animation, the node at index 22, which initially has the value x, has been replaced by $x \cdot k$ where $k \in \{0.00, 0.01, 0.02, ... 10.00\}$. Each of the latter substitutions generates a new function which can be evaluated to produce a single animation frame.

In the AE system used here, the node "tweaks" are performed by passing the base image (and it's embedded genotype) to a custom tool called "evotool". This supports many kinds of manipulations for both the genotype and phenotype. As an example, the following invocation creates a new AE artifact called "out.exp.jpg" by changing the terminal node at index 22 in the file "in.exp.jpg" to "$sin(x)$". The node index 22 was obtained by previous invocation of evotool with the "node list" command sequence which lists terminal nodes.

```
evotool node tweakt default 22 "sin(x)" out.exp.jpg in.exp.jpg
```

Thus the process of animating a function becomes a simple case of scripting the use of evotool to create a sequence of progressively tweaked functions and their phenotypes. The outputted image sequence can then be stitched together using the opensource tool FFmpeg [8], in a manner such as:

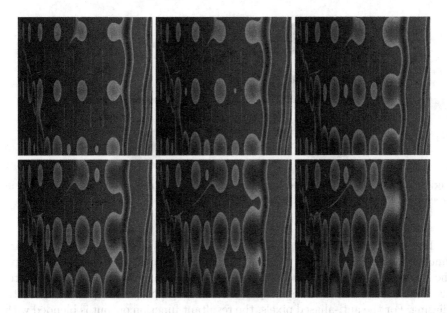

Fig. 2. Six animation frames extracted from a sequence where the terminal at index 22 of the underlying function has assumed the values $x \cdot k$ where $k \in \{0.00, 0.01, 0.02 \ldots 10.00\}$

```
ffmpeg -i out_%d.exp.jpg -c:v libx264 -crf 17 -c:a copy out.mkv
```

To encode the generated image sequence at the default of 25 frames per second as an x264 [9] encoded movie within a Matroska [10] container.

Some experimentation is required to find interesting modulations, as not all nodes contribute constructively to the phenotype[1]. Of course, it is feasible to modulate more than one node simultaneously, and the sequence of tweak inputs can be arbitrarily controlled. For example, the artist may wish to provide a sequence of inputs consistent in magnitude differentials with that of a audio segment to produce animation that follows the sound. Several examples of animated functions, created in this manner, including an audio-modulated examples, are available to view online [11].

4   Mapping of Animated Functions to Typefaces

To produce an animated typeface, each glyph in the typeface must be mapped to an animated function. The first step in this process is to generate mapping images for each glyph. Here, for each glyph, the mapping image is simply that glyph rendered using anti-aliasing in black against a white background at the desired point size. Figure 3 shows such a mapping image along with a mapped function.

---

[1] It is interesting that this "junk DNA" accumulates in AE images, as it apparently does in natural beings.

Fig. 3. An aesthetically evolved image mapped onto the roman alphabet glyph "e" in typeface "Organo" [12] against a black background.

In this case the mapped function has 258 nodes in total of which 100 are function nodes and 158 are terminal nodes. To map the function to the glyph the function is evaluated for each pixel in the map that is not completely white, i.e. those pixels which are either part of the image proper, or part of the anti-aliasing. For the anti-aliased pixels, the resultant function output is blended with the background color in the output image in proportion to the level of grayscale in the map, whereas the ordinary pixels take the function output directly. This approach produces a straightforward silhouette of the evaluated function which accords with the shape of the glyph selected, as is shown in Fig. 3. This is again achieved using "evotool" and the sub-command "map", which takes only three arguments: the output file, the mapping silhouette as a jpg, and the input file to map:

```
evotool map default out.exp.jpg map.jpg in.exp.jpg
```

Thus, each frame from an animated function is mapped onto a glyph. Once this has been done, the resultant image sequence is composited into a movie as described in the previous section. Examples of animated glyphs can also be viewed online [11].

5 Compositing Animated Typefaces into Animated Typescript

By following the process outlined in the previous section for each glyph in a given typeface, a corresponding typeface that is animated is obtained. By combining animated glyphs together into strings, animated typescript can be created, which provides a novel way to animate text.

Whilst there are many ways to composite video together, the method used here is simple. FFmpeg is used to stitch together the chosen glyphs onto a 1080p white background into a single movie in their correct textual positions. An example command is shown below:

```
ffmpeg -i E.mkv -i V.mkv -i O.mkv -i A.mkv -i R.mkv -i T.mkv \
  -filter_complex"
  color=size=1920x1080:c=white [a];
  [0:v] scale=320x320 [E];
  [1:v] scale=320x320 [V];
  [2:v] scale=320x320 [O];
  [3:v] scale=320x320 [A];
  [4:v] scale=320x320 [R];
  [5:v] scale=320x320 [T];
  [a][E] overlay=shortest=1:y=200          [b];
  [b][V] overlay=shortest=1:x=320:y=200    [c];
  [c][O] overlay=shortest=1:x=640:y=200    [d];
  [d][A] overlay=shortest=1:x=960:y=200    [e];
  [e][R] overlay=shortest=1:x=1280:y=200   [f];
  [f][T] overlay=shortest=1:x=1600:y=200   [g]"
```

Whilst this may seem somewhat involved, the invocation is quite straightforward: the input letters are passed to a complex filter that scales and labels each letter, whereupon a series of composite overlays position the scaled input letters accordingly. It is a trivial matter to automate the process of generating such a command sequence for any number of letters, scaling, and relative positioning. Figure 4 shows a single frame from an animation of the text "EVOART" in the font Cantarell-Bold rendered using this process.

Fig. 4. A single frame from animated text "EVOART" in Cantarell-Bold font, where every animation is the same for each letter.

Fig. 5. A single frame from an animated alphabet in Cantarell-Bold font, where every animation is slightly different for each letter.

Fig. 6. A single frame from animated text "EVOART" in Cantarell-Bold font, where every animation is slightly different for each letter.

In the animation frame shown in Fig. 4, all of the glyphs are mapped to the same animation. Whilst this effect may be desired by the artist or designer, it is perhaps more interesting to vary the animation applied to each glyph used in the sequence so that the combined typescript animation has a little more life to it. Figure 6 shows an animation frame taken from constructed alphabet where each glyph has been animated in a different manner to the others. In this case the evaluation range was modulated. Other approaches are to tweak different nodes, or tweak more than one node, or change other evaluation constraints (Fig. 5).

Figure 6 shows an animation frame using this alphabet where the word "EVOART" has been rendered. Fully animated versions of Figs. 4 and 6 can be viewed online [11].

6   Discussion and Conclusion

The ability to animate typefaces with aesthetically pleasing functions is useful for artistic and promotional purposes. It gives text a high quality and lifelike feel that would be painstakingly hard to achieve using manual animation techniques. After the initial cognitive filtering that yields the input functions, the process amends itself to a high degree of automation. Given the limitless supply of diversity available within a given AE system, this entails limitless diversity and customization for animated typefaces and script.

Beyond the immediate artistic merit of animating aesthetically evolved images, at a deeper level there is something potentially very powerful here. An animated function consists of a base function and a sequence of parameterized mutations. The sequence of parameter values is a time series, and thus we can think of an animated function phenotype as being the projection of a time series into a high-dimensional space.

In machine learning (ML), Cover [13] showed that pattern classification problems are more likely to be linearly separable when projected non-linearly into a high dimensional space. This idea is exploited by a wide range of ML techniques from the seemingly artificial support vector machine [14] through to the biology emulating Liquid State Machine neocortex model [15].

Typically, when animating a function as described in this paper, the parameterization sequence is chosen to be a simple geometric sequence with no external meaning, but this need not be so, as was illustrated in the musically driven example shown online [11]. We can extend this idea therefore to *any* time-series data, including those which are bound to meaningful classification and prediction problems.

Ergo, subject to sufficient funding, in future work we plan to explore the potential of AE functions to the application of time-series classification and prediction tasks. The reason this is particularly interesting is that the AE functions have been preselected by an existing form of intelligence, and thus it is interesting to ask to what extent this intelligence is embedded or correlated with the selected functions and their applied computational power. Exploration in this direction might start by evaluating the Lyapunov exponent [16] for some time series for functions taken from increasingly greater generations, i.e. images that are increasingly "aesthetic", to see if there is a correlation between that which is aesthetically interesting and that which is computationally powerful.

The very fact that the created animations have structure, that is, that our brains can distinguish earlier frames from later frames *at all*, indicates clearly that the input information is not destroyed upon projection, strongly implying underlying computational power. This makes the prospect of research in this direction extremely exciting, and we look forward to observing the results.

References

1. Sims, K.: Artificial evolution for computer graphics. Comput. Graph. **25**(4), 319–328 (1991)
2. Todd, S., Latham, W.: Evolutionary Art and Computers. Academic Press, London (1992)
3. Machado, P., Cardoso, A.: NEvAr - the assessment of an evolutionary art tool. In: Proceedings of the AISB 2000 Symposium on Creative and Cultural Aspects and Applications of AI and Cognitive Science, Birmingham, UK (2000)
4. Rooke, S.: Eons of Genetically Evolved Algorithmic Images. Morgan Kaufmann Publishers Inc., San Francisco (2002)
5. McCormack, J.: Aesthetic evolution of l-systems revisited. In: EvoWorkshops, pp. 477–488 (2004)
6. Mills, A.: Evolving aesthetic images. MSc Mini Project Thesis (2005). https://www.ashleymills.com/ae/EvolutionaryArt.pdf
7. Martins, T., Correia, J., Costa, E., Machado, P.: Evotype: evolutionary type design. In: Johnson, C., Carballal, A., Correia, J. (eds.) EvoMUSART 2015. LNCS, vol. 9027, pp. 136–147. Springer, Heidelberg (2015)
8. Ffmpeg. http://www.ffmpeg.org
9. Merritt, L., Vanam, R.: x264: A high performance h. 264/AVC encoder (2006). http://neuron2.net/library/avc/overview_x264_v8_5.pdf
10. Noé, A.: Matroska file format (2009). http://www.matroska.org/files/matroska.pdf
11. Multimedia examples of the artifacts described in this paper. http://www.evoart.club/evomusart2016
12. Nikolov, N.: Organo font landing page. http://logomagazin.com/organo-font/
13. Cover, T.: Geometrical and statistical properties of systems of linear in equalities with applications in pattern recognition. IEEE Trans. Electron. Comput. **3**(EC–14), 326–334 (1965)

14. Cortes, C., Vapnik, V.: Support-vector networks. Mach. Learn. **20**(3), 273–297 (1995)
15. Maass, W., Natschläger, T., Markram, H.: Real-time computing without stable states: a new framework for neural computation based on perturbations. Neural Comput. **14**(11), 2531–2560 (2002)
16. Wolf, A., Swift, J.B., Swinney, H.L., Vastano, J.A.: Determining Lyapunov exponents from a time series. Physica D **16**(3), 285–317 (1985)

An Evolutionary Composer for Real-Time Background Music

R. De Prisco[(⊠)], D. Malandrino, G. Zaccagnino, and R. Zaccagnino

Dipartimento di Informatica, Università di Salerno, 84084 Fisciano, SA, Italy
robdep@unisa.it
http://music.dia.unisa.it

Abstract. Systems for real-time composition of background music respond to changes of the environment by generating music that matches the current state of the environment and/or of the user.

In this paper we propose one such a system that we call EvoBack Music. EvoBackMusic is a multi-agent system that exploits a feed-forward neural network and a multi-objective genetic algorithm to produce background music. The neural network is trained to learn the preferences of the user and such preferences are exploited by the genetic algorithm to compose the music. The composition process takes into account a set of controllers that describe several aspects of the environment, like the dynamism of both the user and the context, other physical characteristics, and the emotional state of the user. Previous system mainly focus on the emotional aspect.

EvoBackMusic has been implemented in Java using Encog and JFugue, and it can be integrated in real and virtual environments.

We have performed several tests to evaluate the system and we report the results of such tests. The tests aimed at analyzing the users' perception about the quality of the produced music compositions.

1 Introduction

Interest for real-time composition of *background music* has grown with the increasing diffusion of Intelligent Virtual Environments, movie music production and interactive media. A system for real-time composition of background music responds to changes of the environment by generating *"appropriate music"* that matches the actual state of the environment and/or of the user.

Background music has an accompanying role that enriches a particular activity or event such as an exhibition of paintings, a theater, the scenes of a film, a Web site, a video game or a virtual environment.

In general, the approach followed by developers, in the game or film industry for example, is to loop pre-composed music tracks which are attached to particular locations or events. One problem with this approach is that users get accustomed to listening the same music over and over again. Another problem is that users associate music to events, by predicting what will be going to happen next. Only recently some commercial sound engines started to include technology that

© Springer International Publishing Switzerland 2016
C. Johnson et al. (Eds.): EvoMUSART 2016, LNCS 9596, pp. 135–151, 2016.
DOI: 10.1007/978-3-319-31008-4_10

allow to increase music diversity. Microsoft DirectMusic[1], for example, provides the possibility to compose several pieces of music which are randomly arranged at run time, but continues to do this in response to some specific events.

The problem of composing real-time background music has attracted considerable attention in the last few years. Most of the existing efforts produce music mainly considering the emotional state of the user, without involving objective aspects of the environment in which the user is immersed.

Related works. The problem of composing background music has been considered by several papers. Downie [1] presents a system that produces Music Creatures following a reactive, behavior-based AI approach. Nakamura et al. [2] implemented a prototype system that generates background music and sound effects for short movies, based on Actors Emotions and Motions. A work done at the University of Edinburgh [3] describes a system that generates atmospheric music in real time to convey fear using suspense and surprise. The system's parameters are controlled by a human director.

Several systems are grounded on research made in the areas of Computer Science and Music Psychology. Systems that control the emotional impact of musical features modify emotionally-relevant structural and performative aspects of music [4–6] by using pre-composed musical scores [5,7] or by making musical composition [8–10]. Most of these systems are built on empirical data obtained from works of psychology [11,12]. Scherer and Zentner [13] established the parameters of influence for the experienced emotion. Meyer [14] analyzed structural characteristics of music and its relation with emotional meaning.

We remark that most of these works only take into account emotional aspects and they use pre-composed music for generating background music. In our approach we investigate the importance of both subjective and objective aspects of an environment for the composition of background music. Furthermore, our system always composes new music, in real-time, according to the musical preferences of the user.

Other papers somewhat related to the subject of automatic music composition and biologically inspired algorithms are [15–18]?.

Contribution of this paper. We provide a new system capable to: *(1)* learn the music preferences of a user, and *(2)* generate background music according to current state the environment and of the user.

From now on, when we talk about environment we implicitly include the user in the environment. In other words the environment includes both subjective aspects, the emotional state of the user, and objectives aspects, the environment in which the user is immersed. Specifically the *subjective aspects* concern the emotional experiences of the user, such as happiness, sadness and so on, while the *objective aspects*, regard the dynamism of the elements in the environment, the density of the elements in the environment, the type of the activity being performed, and a predominant color of the environment.

---

[1] http://www.microsoft.com/DirectX.

In order to compose music we take into account several aspects: harmony, melody, rhythm, and timbre. Thus, we tackle the problem of composing background music with a multi-objective problem. Specifically, we consider: *(1)* the *instrumental objective*, i.e., to find an appropriate instrumental orchestration; *(2)* the *rhythmic objective*, i.e., to find an appropriate rhythmic section; *(3)* the *melodic objective*, i.e., to find good melodic lines; *(4)* the *harmonic objective*, i.e., to find good harmony. Moreover, the problem can be considered as a *dynamic optimization problem* in which the objective functions (i.e., the generated music) change continuously according to the changes of the environment.

From the technical point of view, EvoBackMusic consists of three agents: the Observer Agent, the Composer Agent, and the Sound Agent. The Observer Agent checks for environment state changes and provides such changes to an Artificial Neural Network (ANN). The ANN, trained according to the user preferences, maps the current environment state produced by the environment changes into a *musical state*. Then, the musical state is given to the Composer Agent that, using a multi-objective dynamic genetic algorithm, composes appropriate music. Finally, the Sound Agent converts music compositions into audio.

The three Agents have been implemented in Java. The implementation of the Artificial Neural Network uses the Encog library[2]. The Sound agent communicates with a sequencer implemented using the JFugue library[3].

To evaluate the overall system, we performed several tests that study its behavior and its efficacy. The quality of the produced music, instead, was analyzed by participants recruited for a preliminary evaluation study.

Background. The reader of this paper should be familiar with neural networks and genetic algorithms. Music knowledge will help in understanding the musical aspects. Due to space limitation we will not provide a background section on these topics but we assume that the reader has at least some basic notions about them.

Organization of the paper. In Sect. 2 we provide details about how we encode information from the environment and how we map it into musical information. Then, in Sect. 3 we present the system for producing background music. Finally, Sect. 4 reports the results of tests that we have administered in order to evaluate the system.

2   Environment and Musical States

In this Section we provide details about how we selected the subjective and objective controllers that have to be taken into account when composing background music. We remark that in order to be able to formally model the state and the corresponding music, we had to make some choices, for example about which aspects of the environment to consider, or which instruments to use to produce

---

[2] http://www.heatonresearch.com/encog.
[3] http://www.jfugue.org/.

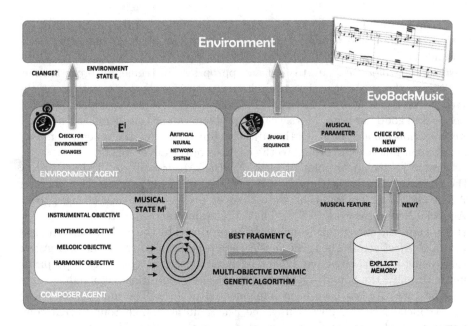

Fig. 1. EvoBackMusic architecture: components and interactions.

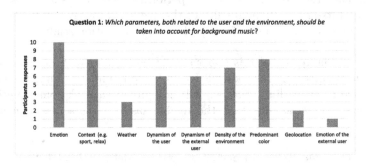

Fig. 2. Environment parameters suggested choices (Question 1).

the music. Clearly these choices are arbitrary and the system can be extended or changed in order to allow more flexibility. Figure 1 provides an overall view of the system. We will provide a description of the components in Sect. 3; however having an idea of what the overall system looks like will be useful to understand the environment and the musical states that we introduce in this section.

2.1 Environment State

The set of parameters needed to describe the environment and the musical state has been chosen considering the results of a test questionnaire administered to a group of 10 music teachers.

Table 1. Controllers.

Feature description	Name	Range
Objective		
Dynamism of the user	e_1	$0, 1, \ldots, 100$
Dynamism of external objects	e_2	$0, 1, \ldots, 100$
Density of the environment	e_3	$0, 1, \ldots, 100$
Context	e_4	$0, 1, \ldots, 5$
Predominant color	e_5	$(x, y, z), 0 \leq, x, y, z \leq 255$
Subjective		
Emotion of the user	e_6	(x, y) coordinate

In Fig. 2 we show a summary of the responses for Question 1. As one can see in the figure, all participants considered the *emotion of the user* an important parameter while just one expressed his propensity to include the *emotion of other people* in the set of the parameters needed to compose background music. We decided to use the parameters that received a clear majority of supporting answers, namely the ones specified in Table 1. We refer to such parameters as *controllers*.

The parameter e_1 indicates how rapidly the user is moving in the environment (0 means a slow user, 100 means a rapidly moving user). The parameter e_2 refers to the speed of other objects available in the environment. The parameter e_3 indicates how dense is the environment, i.e. how many entities, including the user, are present in the environment. The parameter e_4 describes the possible "contexts", such as: fun ($e_4 = 1$), sport ($e_4 = 2$), game ($e_4 = 3$), film ($e_4 = 4$) and relax ($e_4 = 5$). The parameter e_5 describes the predominant color of the environment, in terms of Red-Green-Blue (RGB) components.

Finally, e_6 describes the emotion of the user, represented by using a two-dimensional model of affective experience. Specifically, we used a graphical representation of the circumplex model of affect with the horizontal axis representing the valence dimension (i.e., the intrinsic attractiveness or aversiveness) and the vertical axis representing the arousal (i.e., a state of heightened activity that makes us more alert) or activation dimension [19]. As shown in Fig. 3, for this circular representation that includes 28 words, a greater similarity between two words is represented by their closeness in the space. We enhanced this model with a layered subdivision of the entire space in three subspaces, i.e., a high (serious-like), medium (light-like) and low subspace (quiet-like), according to the intensity of the emotion.

To summarize, we use a *6-tuple* $E^i = [e_1^i, \ldots, e_6^i]$ to describe the state of the environment at the ith time interval. For example $E^{12} = [18, 12, 80, 3, (255, 0, 0), (10, -1)]$ says that at the 12th time interval the user is slowly moving ($e_1 = 18$) and pleased ($e_6 = (10, -1)$), and there is a quite crowded environment ($e_3 = 80$)

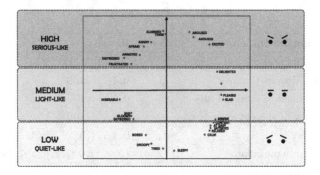

Fig. 3. Circumplex model of affect.

with other entities moving, on average, slower than the user ($e_2 = 12$), in a game context ($e_4 = 3$), and that the predominant color is red ($e_5 = (255, 0, 0)$).

2.2   Musical State

The environment state is built through information obtained in input from the surrounding environment, and it is not used to create music directly. Instead, it is used to compute a musical state.

The set of parameters included in the musical state has been chosen similarly to the set of parameters used for the environment state. Specifically, as shown in Fig. 4, we used the results of the Questions 2 and 3 that we asked to the group of 10 music teachers. Again, the set of parameters that we have included in the musical state, shown in Table 2, were chosen among the choices with a clear majority of supporting answers.

These parameters are grouped into 4 classes: instrumental, rhythmic, harmonic and melodic class.

The instrumental class contains 5 boolean variables $m_1, \ldots, m_5$ that indicate whether the corresponding instrument is used. In the rhythmic class there are three parameters. The first one, m_6, indicates the meter of the music. We consider the following meters: $\frac{2}{4}$ meter ($m_6 = 0$), $\frac{3}{4}$ meter ($m_6 = 1$), $\frac{4}{4}$ meter ($m_6 = 2$),

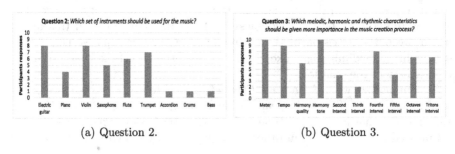

(a) Question 2.   (b) Question 3.

Fig. 4. Participants choices for the musical state.

Table 2. Melodic, harmonic and rhythmic parameters for the musical state.

Feature description	Name	Range
Instrument		
Electric guitar	m_1	Binary value
Violin	m_2	Binary value
Saxophone	m_3	Binary value
Flute	m_4	Binary value
Trumpet	m_5	Binary value
Rhythm		
Meter	m_6	$0, 1, \ldots, 5$
Tempo	m_7	$40, 41, \ldots, 208$
Harmony		
Quality	m_8	Binary value
Melody		
Melodic thirds	m_9	Integer
Melodic fifths	m_{10}	Integer
Melodic octaves	m_{11}	Integer
Melodic tritons	m_{12}	Integer

$\frac{6}{8}$ meter ($m_6 = 3$), $\frac{9}{8}$ meter ($m_6 = 4$), $\frac{12}{8}$ meter ($m_6 = 5$). The second parameter m_7 gives the velocity of the execution (tempo) ranging from *Grave* ($m_7 = 40$), that is very slow, to *Prestissimo* ($m_7 = 208$), that is very quick.

The harmonic class contains only one parameter, m_8 which is a boolean variable that indicates whether the music is in a minor tonality ($m_8 = 0$) or in a major tonality ($m_8 = 1$).

Finally in the melodic class we have four parameters, $m_9, \ldots, m_{12}$, which provide some information about the melodic lines. These parameters give the number of jumps of thirds, fifths, octaves and tritons in the melody, respectively.

To summarize, a musical state can be described with a 12-tuple $[m_1^i, \ldots, m_{12}^i]$. As an example, the musical state $M^{15} = (0, 1, 0, 0, 1, 2, 90, 8, 5, 6, 0)$ for the 15th time interval says that for such an interval the Composer agent should use orchestral strings and a trumpet ($(m_1, m_2, m_3, m_4, m_5) = (0, 1, 0, 0, 1)$), the meter of the music should be 4/4 ($m_6 = 2$), the tempo 90 beats per minute ($m_7 = 90$). The total melodic jumps to be used are 8,5,6,1 for, respectively, thirds, fifths, octaves and tritones ($(m_9, m_{10}, m_{11}, m_{12}) = (8, 5, 6, 1)$).

3   EvoBackMusic Components

In this Section we describe EvoBackMusic. EvoBackMusic consists of three agents: the *Observer Agent*, the *Composer Agent* and the *Sound Agent*. The Observer agent interacts with the environment getting input from it, and, after an appropriate transformation, relays such input to the Composer Agent. The Composer Agent is responsible of the music creation. The created music is passed to the Sound Agent which simply plays it and the music can be heard in the environment. Figure 1, shown in the previous section, provides an overall view of the system.

3.1   The Observer Agent

The Observer Agent checks the state of the environment looking for changes with respect to the latest update, at fixed intervals of time. We consider interval of times of a fixed length, say T seconds, where T is a fixed global parameter of the system. We will denote with E^i the state of the environment at the beginning of the ith time interval, that goes from $(i-1) \cdot T$ to $i \cdot T$ seconds, starting with the first time interval I^1.

Once obtained the environment state the Observer Agent maps it into a Musical State. In order to do so the Observer exploits an artificial neural network. Such a network must be trained in advance according to the musical preferences of the user. This means that the user, in a preliminary training phase, acts has an implicit composer for the music that he will hear later on.

The artificial neural network is a fully connected three-layer feed-forward neural network with sigmoidal activation function. The input layer models the environment, and therefore contains 9 neurons: e_1, e_2, e_3, e_4, $e_5(x)$, $e_5(y)$, $e_5(z)$, $e_6(x)$, $e_6(y)$, where $e_5(x)$, $e_5(y)$, $e_5(z)$ are the x, y, z coordinates of e_5, respectively, and $e_6(x)$, $e_6(y)$ are the x, y coordinates of e_6 (see the Sect. 2.1 to recall the meaning of the variables). The output layer represents a musical state and, therefore, contains 12 neurons: $m_1, \ldots, m_{12}$ (see the Sect. 2.2 to recall the meaning of the variables). We recall that each neuron (both input and output) represents a value normalized between 0 and 1.

The number of neurons in the hidden layer is a critical choice, since an high number could wrongly involve more equations than free variables, resulting in a system with a less generalization ability, while a low number could imply a system with less learning ability and, therefore less robust. Also the learning rate and the momentum are crucial parameters of the neural network.

In order to fine tune the network structure we performed a training phase. For such a training phase we fixed a data set of 50 pairs of environment states and musical states. For each possible choice of the number of neurons in the hidden layer (10 random choices), value of the learning rate and the value of the momentum (9 possible values in a range between 0 and 1), we computed 10 times the training error by running the neural network for 100.000 epochs. Then for each of the above 810 experiments, each repeated 10 times, we computed the average training error. Finally we ranked all the 810 cases based on the average training error. Table 3 shows the best average training errors.

As we can see, the lowest training error is obtained by a network with 18 neurons in the hidden layer, a learning rate of 0.7 and a momentum of 0.5. Hence we choose such parameters as default values of the neural network used by EvoBackMusic. Once the network has been trained its use is straightforward:

Table 3. Ranking of the training errors.

#	Error	Parameters	#	Error	Parameters	#	Error	Parameters
1	0.042	(18,0.7,0.5)	16	0.098	(9,0.5,0.7)	31	0.157	(18,0.5,0.7)
2	0.059	(54,0.7,0.3)	17	0.099	(36,0.5,0.7)	32	0.211	(21,0.7,0.3)
3	0.065	(41,0.7,0.7)	18	0.102	(18,0.7,0.7)	33	0.214	(9,0.7,0.5)
4	0.065	(45,0.5,0.7)	19	0.102	(13,0.5,0.7)	34	0.217	(13,0.7,0.5)
5	0.065	(54,0.5,0.7)	20	0.103	(41,0.7,0.5)	35	0.221	(36,0.7,0.5)
6	0.066	(41,0.5,0.7)	21	0.103	(30,0.7,0.3)	36	0.223	(30,0.5,0.7)
7	0.069	(13,0.5,0.5)	22	0.112	(54,0.7,0.7)	37	0.235	(36,0.7,0.7)
8	0.072	(45,0.7,0.7)	23	0.113	(45,0.7,0.5)	38	0.242	(54,0.7,0.5)
9	0.084	(13,0.7,0.3)	24	0.121	(30,0.7,0.7)	39	0.242	(36,0.7,0.3)
10	0.088	(21,0.7,0.7)	25	0.121	(27,0.5,0.7)	40	0.242	(21,0.5,0.5)
11	0.088	(27,0.7,0.7)	26	0.122	(18,0.7,0.3)	41	0.312	(30,0.7,0.5)
12	0.092	(21,0.5,0.7)	27	0.123	(9,0.5,0.5)	42	0.322	(27,0.7,0.3)
13	0.093	(9,0.7,0.7)	28	0.132	(18,0.5,0.5)	43	0.324	(9,0.7,0.3)
14	0.098	(21,0.7,0.5)	29	0.141	(13,0.7,0.7)	44	0.342	(27,0.5,0.5)
15	0.098	(27,0.7,0.5)	30	0.145	(45,0.7,0.3)	45	0.347	(41,0.7,0.3)

one gives the 9-value input representation of the environment state and the network produces the 13-value output that represents a musical state.

3.2 The Composer Agent

The Composer agent, given as input a musical state, composes a fragment of music lasting T seconds, using a dynamic multi-objective genetic algorithm. For each time interval I^i the algorithm evolves one generation and then creates a new music fragment according to the current musical state M^i.

The multi-objective genetic algorithm tries to compose "good" musical fragments. As described in Sect. 1, it considers the following objectives: the instrumental, the melodic, the harmonic, and the rhythmic objective functions. We use an *explicit memory* approach in order to continuously adapt the solution to a changing environment, reusing the information gained in the past. In the following we provide the details of the GA used by the Composer Agent.

Chromosome and Gene Representation. The population is composed of individuals that are fragments of music. Each fragment lasts for T seconds which allows us to place n beats (the exact value of n depends on the actual value of T and the speed of execution of the music, which is a parameter in the musical state). So, each chromosome represents a fragment of music of duration T, and each gene of the chromosome represents a beat of the fragment.

We represent a chromosome (individual) of n genes as an array C of dimension $14 \times n$ where we store information related to the music fragment. For each beat j, the vector column at position j is a *gene*. We denote a chromosome C with $C = C_1, \ldots, C_n$ where C_i is the gene at beat i. Each gene C_i contains 14 values. Table 4 provides a summary of the music information parameters.

Table 4. Gene Representation

Feature description	Name	Range
Instrument layer		
Electric guitar activation	c_1	Binary value
Violin activation	c_2	Binary value
Saxophone activation	c_3	Binary value
Flute activation	c_4	Binary value
Trumpet activation	c_5	Binary value
Rhythm layer		
Meter	c_6	$0, 1, \ldots, 5$
Tempo	c_7	$40, 41, \ldots, 208$
Melody layer		
Electrical guitar notes	c_8	Array of *Note*
Violin notes	c_9	Array of *Note*
Saxophone notes	c_{10}	Array of *Note*
Flute notes	c_{11}	Array of *Note*
Trumpet notes	c_{12}	Array of *Note*
Harmony layer		
Quality	c_{13}	Binary value
Tone	c_{14}	$0, 1, \ldots, 11$

Musical information in each gene are grouped into 4 classes: Instrumental, Rhythmic, Melodic, and Harmonic classes. The Instrumental class contains 5 boolean variables $c_1, \ldots, c_5$ that indicate whether the corresponding instrument is used. In the Rhythmic class there are three parameters. The first one, i.e., c_6, indicates the meter of the music: $\frac{2}{4}$ m ($c_6 = 0$), $\frac{3}{4}$ m ($c_6 = 1$), $\frac{4}{4}$ m ($c_6 = 2$), $\frac{6}{8}$ m ($c_6 = 3$), $\frac{9}{8}$ m ($c_6 = 4$), $\frac{12}{8}$ m ($c_6 = 5$). The second parameter c_7 gives the velocity of the execution ranging from *Grave* ($c_7 = 40$), that is very slow, to *Prestissimo* ($c_7 = 208$), that is very quick.

The Melodic layer contains the sequence of notes for each instrument in the instrument layer. Each note is represented as a triple *midi-value, duration, ligature*: midi-value is an integer between 21 and 108 representing the midi value of the note. Note that each type of instrument has a range of possible notes that we set as parameters. Therefore, we require that for each type of instrument, the midi value of a note has to range between the minimum and maximum value for the instrument itself. Duration is 0 if the duration of the note is $\frac{4}{4}$, 1 if the duration of the note is $\frac{2}{4}$, 2 if the duration of the note is $\frac{1}{4}$, 3 if the duration of the note is $\frac{1}{8}$, 4 if the duration of the note is $\frac{1}{16}$, 5 if the duration of the note is $\frac{1}{32}$ and 6 if the duration of the note is $\frac{1}{64}$; ligature is a binary value such that 1 means that the note is ligate to the next note, 0 otherwise. The ligature is necessary in order to represents note whose duration exceeds the duration of a single beat. Notice that, although we could have use a representation that does not need ligatures by simply using a duration equivalent to the total duration of the ligated notes, we preferred to have at least one note in each beat because this ease the manipulation of genes.

The Harmonic layer contains the parameter c_{13} which is a boolean variable that indicates whether the music is in a minor tonality ($c_{13} = 0$) or in a major tonality ($c_{13} = 1$) and the parameter c_{14} which is an integer that indicates the current tone, ranging from 0 to represent C up to 11 to represent B (with 1 representing $C\#$, 2 representing D, and so on).

Initial Population. In the very first iteration, we start from a random initial population of K individuals, where $K = 50$ is a fixed parameter. Such a population is built by selecting a random gene for each beat. In subsequent iterations, the population is half random and half chosen among the solutions of the last iteration before the environment change. This is motivated by somehow tying the previous music to the current one so that to have a smooth passage from the music corresponding to the previous environment to that corresponding to the current environment.

Evaluation Function. The algorithm uses 4 objective functions: an instrumental objective function f_I, a rhythmic objective function f_R, a melodic objective function f_M and an harmonic objective function f_H.

Given a chromosome C and the current musical state M we compute: $f_I(C)$ by defining the *instrumental distance* between C and M, $f_R(C)$ by defining the *rhythmic distance* between C and M, $f_M(C)$ by defining the *melodic distance* between C and M and $f_H(C)$ by defining the *harmonic distance* between C and M. They represent measures of similarity (the lower the distance, the similar the states), each of them with respect to one of the classes of parameters.

The *instrumental distance* between C and M is $f_I(C) = \sum_{k=1}^{5} |m_k - \frac{\sum_{j=1}^{n} c_k^j}{n}|$. The *rhythmic distance* is $f_R(C) = \sum_{k=6}^{7} |m_k - \frac{\sum_{j=1}^{n} c_k^j}{n}|$. The *harmonic distance* is $f_h(C) = |m_8 - \frac{\sum_{j=1}^{n} c_8^j}{n}|$ (the formula for the *harmonic distance* does not have the external summation because there is only one harmonic musical feature).

To define the *melodic distance* recall that gene C_i contains 5 melodic fragments for beat i and that there are n genes for $i = 1, \ldots, n$. For each $i = 1, \ldots, n$, let $c_{i,j}$ be the melodic fragment for beat i and for the instrument j, with $j = 8, \ldots 11$, and let $c_{i,j}^3$, $c_{i,j}^5$, $c_{i,j}^8$ and $c_{i,j}^T$ the number of melodic thirds, fifths, octaves and tritons in $c_{i,j}$ respectively. The *melodic distance* between C and M is:

$$f_m(C) = |(m_9 + m_9 + m_{10} + m_{11})$$
$$- \sum_{i=1}^{n} \frac{(\sum_{j=1}^{5} c_{i,j}^3) + (\sum_{j=1}^{5} c_{i,j}^5) + (\sum_{j=1}^{5} c_{i,j}^8) + (\sum_{j=1}^{5} c_{i,j}^T)}{5n}|$$

The objective is to minimize each of the evaluation functions, that is we want to get a chromosome that is as close as possible to the originating musical state.

Evolution Operators. In order to evolve the population, we apply *mutation* and *crossover* operators. Moreover, we used the elitism technique to improve the individuals in the current population by keeping an *external population* P_e that selects chromosomes that have at least one of the following characteristics: *(1)* genes with the same harmonic features, *(2)* genes with the same instruments activated, and *(3)* genes with the same rhythmic features.

In each evolution step we apply the elitism operator to update P_e, by inserting new elements that satisfy the above criteria; we then choose the best K individuals to keep the external population at size K. Note that the elitism operator does not produce new individuals; it only preserves some individuals in the hope that future generations will inherit some of their characteristics.

We defined the *production population* P_p as the set consisting of K chromosomes from the current population and K chromosomes from the external population P_e. Moreover we also included in P_p the individuals that have been recently given in output; these individuals are kept in an *explicit memory*; this strategy has been suggested in [20]. The crossover and mutation operators produce new individuals starting from the ones in P_p.

- **Classic crossover.** This operator works like a standard crossover operator. Given two chromosomes $C^1, C^2 \in P_p$, it selects an index j randomly, and generates the chromosome $C^3 = C_1^1, \ldots, C_j^1, C_{j+1}^2, \ldots, C_N^2$.
- **Classic mutation.** This operator creates new chromosomes starting from a chromosome of P_p. $\forall C = C_1, \ldots, C_n \in P_p$, selects a random $j \in [1, n]$ and replaces the gene C_j with a random gene C_j'.
- **Emotion-Harmony mutation.** This operator acts on music quality (minor or major). Let E the actual environment state, $\forall C = C_1, \ldots, C_n \in P_p$, if $e_6(x) \geq 0$ (positive emotion) then for the gene C_j' set $c_{14} = 1$ else $c_{14} = 0$.
- **Color-Instruments mutation.** This operator acts on the total active instruments using the number of light/dark colors. Let E the actual environment state, $\forall C = C_1, \ldots, C_n \in P_p$ and let $e_6(x), e_6(y), e_6(z)$ the predominant color RGB components. Let $\hat{e}_6$ be the average of the three components, then:
  - if $0 \leq \hat{e}_6 < 80$ then $\forall C_j'$ we set $c_3 = c_4 = 1$ and $c_1 = c_2 = c_5 = 0$.
  - if $80 \leq \hat{e}_6 \leq 150$ then $\forall C_j'$ we set $c_1 = c_2 = c_5 = 1$ and $c_3 = c_4 = 0$.
  - if $150 < \hat{e}_6 \leq 255$ then $\forall C_j'$ we set $c_1 = c_2 = c_3 = c_4 = c_5 = 1$.

- **Dynamism-Frequency mutation.** This operator acts on the dynamism of the environment. Let $E = e_1, e_2, \ldots, e_6$ the current environment state, we consider the dynamism information e_1 and e_2 and let $\hat{e} = (e_1 + e_2)/2$. $\forall C = C_1, \ldots, C_n \in P_p$, if $\hat{e} \leq 50$ (slow average dynamism), then $\forall$ gene C'_j, and for each sequence of notes, we first divide the sequence in two halves of equal length, possibly splitting the middle note, then we delete the second half of the sequence and we double the duration of the notes in the first half. Conversely, if $\hat{e} > 50$ we split each note of the sequence into two notes of duration equal to half of the duration of the original note.

For each generation the Composer Agent selects the best individual in the current population and writes it in P_e. Since we use a multi-objective fitness function the best individuals are those in the Pareto's front, that is the set of all non-dominated chromosomes. To have a single output we choose the rhythmic function to select the best chromosome in the Pareto's front.

Notice that, depending on the actual value of T, the evolution algorithm can produce several outputs for each time interval, since it needs less than t seconds to generate a single output. Since the Sound agent only needs one music fragment for each time interval, it will simply use only of the produced outputs.

Music fragments are purged from the explicit memory after they become old enough: all the outputs relative to the last Z time intervals are kept in the memory (we used $Z = 5$), while all older the outputs are deleted.

3.3 The Sound Agent

The Sound Agent uses a JFugue sequencer to generate audio by using the fragments generated by the Composer Agent. Whenever the Sound Agent needs a new music fragment, that is, at the beginning of a new time interval, it looks in the external memory and reads the first music fragment which has been given in output for the current time interval (eventually skipping older music fragments). Then, it converts the fragment to audio by mapping each fragment information (notes, rhythm and so on) into JFugue instructions (Patterns, Players and so on). The Sound agents adds also a bass and drum accompaniment by exploiting patterns from a pre-compiled library. Each specific pattern is chosen on the basis of the harmonic and rhythmic information contained in the music fragment.

4   Evaluation Study

Judging the music that a system produces is very difficult, since this is a uncertain process, heavily relying on personal evaluations. In fact, in [9,21] authors stressed the concept of subjective vision about both perceptions and emotions induced by music. They also state that music is inherently subjective and there is a strong correlation between what is seen and what is heard. Most of the works analyzed individuals' feelings in terms of impression adjectives (i.e., happy vs. sad, heartrending vs. not heartrending, and so on) [22], or subjective evaluations [23]. Finally, self-reported data has been often used to evaluate music-related systems/software [4,8].

To evaluate EvoBackMusic we performed a subjective evaluation. Our goal was not to quantify and explain how the produced music influences the individuals' emotional states, but that of establishing whether the user is satisfied by the change in the produced music as a result of changes in the environment, which includes his/her emotional state.

4.1 Methodology

For our preliminary evaluation study we recruited 20 people amongst musicians and people interested in music but with little experience in music composition. We emphasize that participation in the study was voluntary and anonymous, and participants were not compensated for taking part in the interviews.

The study consisted of three phases (see [24]), namely: *(1)* a *Preliminary phase* in which we administered a preliminary survey questionnaire, *(2)* a *Testing phase*, in which we asked participants to perform specific tasks, and finally, *(3)* a *Summary phase*, in which we asked our sample to fill out a summary questionnaire. All questionnaires are publicly available online[4].

The goal of the preliminary survey questionnaire was: to collect demographic information (i.e., age, education level), to get information about the participant's knowledge and attitude toward music, and to ask specific questions about EvoBackMusic (e.g., context for which it would be useful).

In the *Testing phase* we asked participants to create a dataset of 40 pairs of (environment states, MIDI file), by repeating the following task for 40 times:

1. Using the EvoBackMusic graphical interface, select an environment state according to some indications.
2. Select a MIDI file from a pre-defined set of MIDI files with music that the user judges appropriate for the selected environment state (this assumed that the user was familiar with the pre-defined set of MIDI file).

Then the user trained the artificial neural network using the EvoBackMusic graphical interface. Notice that although the default network uses the best parameters (number of neurons in the hidden layer, learning rate, momentum) that we have found in the training phase of the neural network, the system gives the user the possibility of changing these parameters at run time.

At the end of the *Testing phase*, the user evaluates (through a test questionnaire) EvoBackMusic by listening to the music it produces for the environment states in the dataset previously created.

Finally, in the *Summary phase*, we administered the summary survey questionnaire whose goal was to detect whether any change has occurred in the users' opinions after gaining a greater consciousness about the potentialities of the proposed system. The summary questionnaire consisted of three questions that were already asked in the preliminary questionnaire plus one final question about the output of EvoBackMusic.

---

[4] In the download page of the site http://music.di.unisa.it.

4.2 Results

The results of the preliminary questionnaire showed that most participants (80 %) were keen on music (with 60 % very keen on music), while only few (20 %) were neutral. The age ranged from 30 to 50 years, and the education level included 30 % with a bachelor degree, 30 % with a master degree and 40 % with a conservatory degree. Finally, 60 % of the participants were acquainted with computers and computer music while the remaining 40 % was not.

Results of the *Testing phase* showed that EvoBackMusic was very easy to use (see Fig. 5(a)), and that it was able to generate appropriate music, really matching the users' expectations (See Fig. 5(b)).

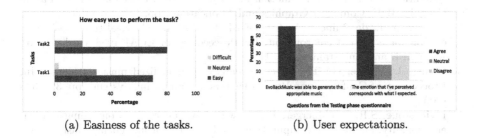

(a) Easiness of the tasks.   (b) User expectations.

Fig. 5. (a) 5-Point Likert scores, (b) Percentages of agreement/disagreement (Color figure online).

The analysis show that the music produced by EvoBackMusic met the expectations of the participants. We recall that the with the summary survey questionnaire we performed a comparative assessment between some specific questions asked before and after the testing phase, in order to evaluate if changes occurred in users' opinions about music importance, general knowledge about the addressed topic and the most suitable activities for which could be useful to use EvoBackMusic. About the last question, three activities, that is "Relax", "Work" and "Game", were selected more than the previous phase (percentage increase of 14 %, 75 %, and 25 %, respectively). EvoBackMusic was also able to increase interest in music and in background music composition. Finally, at the question: *"Overall, I am satisfied with how this system produced music that matches my preferences"*, 80 % of participants expressed their satisfaction.

5 Conclusions

In this paper we have presented a real-time background music composer. The system takes into consideration many parameters in order to guide the composition process. We have implemented and tested the overall system. A video file with a simulation of the Pacman game, integrated into EvoBackMusic, can be found online (In the download page of the site http://music.di.unisa.it). By using

EvoBackMusic (previously trained) we change the states of the game, obtaining a background music for each of such changes. Future works include on one hand a more extensive test phase in order to validate the choice of the controllers that describe the environment and on the other hand the integration of the system into real or virtual environment (games, movies, augmented reality, and so on).

References

1. Downie, M.: Behavior animation, music: the music and movement of synthetic characters. MIT (2011)
2. Nakamura, J., Kaku, T., Noma, T., Yoshida, S.: Automating background music generation based on actors' emotion and motions. In: Proceedings of the Pacific Graphics 1993, Computer Graphics and Applications, pp. 147–161 (1993)
3. Robertson, J., de Quincey, A., Stapleford, T., Wiggins, G.: Real-time music generation for a virtual environment. In: Proceedings of ECAI 1998 Workshop on AI/ALIFE and Entartainment (1998)
4. Chung, J., Vercoe, G.: The affective remixer: personalized music arranging. In: Proceedings of Conference on Human Factors in Computing Systems, pp. 393–398 (2006)
5. Livingstone, S.R., Muhlberger, R., Brown, A.R., Loch, A.: Controlling musical emotionality: an affective computational architecture for influencing musical emotions. Digit. Creativity 18(1), 43–53 (2007)
6. Winter, R.: Interactive music: compositional techniques for communicating different emotional qualities. Master thesis, University of New York, NY (2005)
7. López, A.R., Oliveira, A.P., Cardoso, A.: Real-time emotion-driven music engine. In: Proceedings of International Conference on Computational Creativity (2010)
8. Kim, S., André, E.: Composing affective music with a generate and sense approach. In: Barr, V., Markov, Z. (eds.) FLAIRS Conference, pp. 38–43. AAAI Press (2004)
9. Casella, P., Paiva, A.C.R.: MAgentA: an architecture for real time automatic composition of background music. In: de Antonio, A., Aylett, R.S., Ballin, D. (eds.) IVA 2001. LNCS (LNAI), vol. 2190, p. 224. Springer, Heidelberg (2001)
10. Wassermann, K.C., Eng, K., Verschure, P.F.M.J.: Live soundscape composition based on synthetic emotions. IEEE MultiMed. 10(4), 82–90 (2003)
11. Gabrielsson, A., Lindstrom, E.: The Influence of Musical Structure on Emotional Expression. In: Juslin, P., Sloboda, J.A. (eds.) Music and Emotion: Theory and Research. Oxford University Press, Oxford (2001)
12. Schubert, E.: Measurement and time series analysis of emotion in music. Ph.D. thesis, University of New South Wales, New South Wales (1999)
13. Scherer, K., Zentner, M.R.: Emotional effects of music: production rules. In: Juslin, P., Sloboda, J.A. (eds.) Music and Emotion: Theory and Research. Oxford University Press, Oxford (2001)
14. Meyer, L.: Emotion and Meaning in Music. University of Chicago Press, Chicago (1956)
15. Bilotta, E., Pantano, P.: Artificial life music tells of complexity. In: ALMMA 2001: Proceedings of the Workshop on Artificial Life Models for Musical Applications, pp. 17–28 (2001)
16. Bilotta, E., Pantano, P.: Synthetic harmonies: an approach to musical semiosis by means of cellular automata. Leonardo 35(2), 153–159 (2002)

17. De Prisco, R., Zaccagnino, G., Zaccagnino, R.: Evobasscomposer: a multi-objective genetic algorithm for 4-voice compositions. In: Genetic and Evolutionary Computation Conference, GECCO 2010, Proceedings, Portland, Oregon, USA, 7–11 July 2010, pp. 817–818 (2010)
18. De Prisco, R., Zaccagnino, R.: An evolutionary music composer algorithm for bass harmonization. In: Giacobini, M., et al. (eds.) EvoWorkshops 2009. LNCS, vol. 5484, pp. 567–572. Springer, Heidelberg (2009)
19. Russell, J.: A circumplex model of affect. J. Pers. Soc. Psychol. **39**(6), 1161–1178 (1980)
20. Mori, N., Imanishi, S., Kita, H., Nishikawa, Y.: Adaptation to changing environments by means of the memory based thermodynamical genetic algorithm. In: ICGA 1997, East Lansing, MI, pp. 299–306 (1997)
21. Boenn, G., Brain, M., De Vos, M., ffitch, J.: Automatic composition of melodic and harmonic music by answer set programming. In: Garcia de la Banda, M., Pontelli, E. (eds.) ICLP 2008. LNCS, vol. 5366, pp. 160–174. Springer, Heidelberg (2008)
22. Legaspi, R., Hashimoto, Y., Moriyama, K., Kurihara, S., Numao, M.: Music compositional intelligence with an affective flavor. In: IUI, pp. 216–224 (2007)
23. Unehara, M., Onisawa, T.: Music composition system based on subjective evaluation. In: SMC, pp. 980–986. IEEE (2003)
24. Malandrino, D., Scarano, V., Spinelli, R.: How increased awareness can impact attitudes and behaviors toward online privacy protection. In: SocialCom, pp. 57–62. IEEE (2013)

Iterative Brush Path Extraction Algorithm for Aiding Flock Brush Simulation of Stroke-Based Painterly Rendering

Tieta Putri[✉] and Ramakrishnan Mukundan

University of Canterbury, Christchurch, New Zealand
tieta.putri@pg.canterbury.ac.nz

Abstract. Painterly algorithms form an important part of non-photorealistic rendering (NPR) techniques where the primary aim is to incorporate expressive and stylistic qualities in the output. Extraction, representation and analysis of brush stroke parameters are essential for mapping artistic styles in stroke based rendering (SBR) applications. In this paper, we present a novel iterative method for extracting brush stroke regions and paths for aiding a particle swarm based SBR process. The algorithm and its implementation aspects are discussed in detail. Experimental results are presented showing the painterly rendering of input images and the extracted brush paths.

Keywords: Computational intelligence · Non-photorealistic rendering · Brush stroke extraction · Painterly rendering · Flock simulation · Autonomous agents · Swarm intelligence

1 Introduction

Non-photorealistic rendering is a computer graphics research area that focuses on the generation of output with different styles of art [1]. Some examples of artistic styles that are frequently used in NPR are: cartoonization, pen-and-ink illustration, watercolour illustration, geometric shapes rendering, stippling and painterly rendering. Painterly rendering algorithms try to extract, represent, analyze or generate painting styles. As a field with direct applications in key areas like computer graphics, artificial intelligence, digital image processing, and digital art; a lot of research and development have gone into painterly rendering methods.

The existence of many painting styles with each of them having several unique characteristics has motivated considerable research into the characteristics of painting styles, particularly brush stroke models and colour palette characteristics. According to Gooch and Gooch [1], there are two approaches for simulating artistic media. The first one is to simulate the physics aspect of work of art, and the second one is to emulate the appearance of a chosen medium. The second approach is the most popular due to the applications of a wide range of image processing techniques that can be explored. This paper used the second approach.

© Springer International Publishing Switzerland 2016
C. Johnson et al. (Eds.): EvoMUSART 2016, LNCS 9596, pp. 152–162, 2016.
DOI: 10.1007/978-3-319-31008-4_11

The work in this paper is related to stroke-based rendering (SBR), which has the aim of observing and reproducing the hand-painted appearance created by artists. For a comprehensive survey of SBR algorithms, please refer to [2]. Brush stroke extraction finds several applications in painterly rendering as well as the analysis and classification of paintings. Figure 1, gives a scheme of the processes where brush stroke extraction algorithms are generally applied.

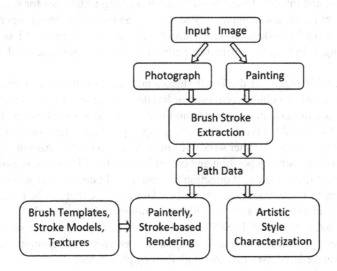

Fig. 1. Schematic showing the application of brush stroke extraction in painterly rendering and analysis.

The paper presents a novel algorithm for automatic computation of brush stroke regions and for the extraction of path data, specifically for producing a painterly rendering of photographic images. The algorithm could also form a basis for more complex methods for brush stroke extraction with paintings as input. The iterative algorithm uses a circular blob that can shrink or expand to match the irregularity of the strokes. The algorithm is designed so that the extracted path closely matches with the brush regions of homogeneous colour. The SBR will be done using the path data that contains the computed radius and the reference colour of the circular blobs at each position. The path will be used as guidance for the brush bristles to move throughout the canvas. The brush bristles are going to be simulated as an autonomous flock of particles which are going to steer themselves in accordance to the given path data.

This paper is organized as follows: In Sect. 2, we take a look at some related work in stroke-based rendering. Section 3 will give a detailed description of the algorithm and discuss important implementation related aspects. Then in Sect. 4 we provide some results and discussions. Finally, Sect. 5 summarizes the work presented in the paper and outlines future directions.

2   Related Works

2.1   Stroke-Based Rendering Techniques

One of the early research works in the area of stroke-based rendering was done by Whitted [3], using anti-aliasing methods for drawing straight and curved brush strokes. Strassman [4] and Pham [5] used brush stroke modelling techniques for simulating the traditional sumi-e Japanese painting. Pudet [6] explored curve-fitting algorithm with Bezier curves for hand-sketched brush strokes. Hertzmann [7] proposed an algorithm using varying brush sizes, where the paths of curved strokes are modelled using B-splines.

Strassman [4] identified the brush strokes simulation process as a union of four components: brush, an object composed of bristles; stroke, a list of brush position and widths; dip, the initial state of a brush; and paint, a rendering method for the strokes. Huang et al. [8] consider a brush stroke as a group of paint drops that move in a cohesive manner over the canvas. In other words, it can be seen as a flock of painting agents with each of the agents carrying position and colour information. The flock is moved across the canvas with the guidance of direction information obtained from the input image.

According to Hertzmann [2], a stroke is a data structure that can be rendered in the image plane and a stipple is a stroke with two parameters: its position in cartesian coordinate (x, y) and the radius r. In SBR, images are created by combining strokes into an image structure, a data structure containing: a canvas with a background colour and/or texture; and an ordered list of strokes with defined parameters.

2.2   Brush Stroke Extraction

As previously mentioned, brush stroke extraction has an important role in the area of digital painting analysis. Li et al. [9] described a brush stroke extraction for distinguishing Van Gogh's paintings from his contemporaries. They also apply the extraction for distinguishing Van Gogh's paintings from two different periods, which are Paris and Arles – St. Remy period. Their work consists of developing: statistical framework for the assessment of the distinction level of different painting categories; brush stroke extraction algorithm using edge detection and segmentation; and numerical features for brush stroke characterization. They used edge detection algorithm called EDISON edge detection algorithm developed by Meer and Georgescu [10]. After edges are detected, edge linking algorithm and enclosing operation are performed in order to close the gaps between edge segments. Then, the processed edges are extracted using the connected component labelling. Finally, brush stroke conditions are defined as: the brush skeleton not severely branched; the ratio of broadness to length is within the range of [0.05, 1.0]; and the ratio of the brush size to two times length times width span is within [0.5, 2.0]. The brush skeleton is produced by the thinning operation of the extracted connected components. The result of their algorithm can be seen of Fig. 2.

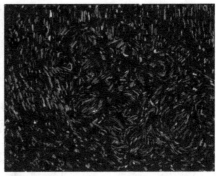

Fig. 2. The result of Li's brush stroke extraction algorithm applied to Van Gogh's painting "Cabbages and Onions" [9].

Johnson et al. [11] did a mathematical analysis for the classification of Van Gogh paintings. They examine high resolution greyscale scans of 101 paintings, which consist of: 82 paintings by Van Gogh, 6 paintings by other painters and 13 others which are still questioned to be Van Gogh or non-Van Gogh by art experts. In their research, they use combine two kinds of features that are extracted from the paintings, which are texture-based feature obtained by wavelets and stroke-based geometric features obtained by edge detection. They argue that "it is extremely challenging to locate strokes accurately from grey-scale images in a fully automated manner" [11].

Berezhnoy et al. [12] elaborated a method called as POET (prevailing orientation extraction technique). This method focuses on brush stroke texture orientation extraction for segmenting individual brush strokes in Van Gogh's painting. The method consists of two stages: the filtering stage and the orientation extraction stage. In the filtering stage, a circular filter which has invariant for rotation and good response for band-passing is applied. A set of binary oriented objects whose properties are used to extracted the orientation. The evaluation of POET is based on the cross comparison between the judgments of POET and human subjects.

Our brush stroke path extraction algorithm is based on the previously developed method by Seo and Lee [13]. Their method is the extension of Gooch's method in [14]. In their work, Seo and Lee described a method that uses Maximum Homogeneity Neighbour (MHN) filter in order to make the region extraction more precise. After applying MHN filter and extracting all of the regions, they smooth the region by using morphological operations such as hole filling, dilation and erosion. Then, medial lines are obtained from the morphed regions using least square approximation. Finally, new brush strokes are generated by inflating the medial lines to their respective region boundary (see Fig. 3).

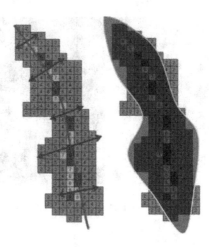

Fig. 3. Stroke generation by inflating the medial line extracted from a brush stroke region [13].

2.3   Flock Brush Simulation

Swarm intelligence (SI) is a technique in artificial intelligence that is based on the study of collective behaviour in decentralized, self-organized systems [19]. The system is usually constructed from a population of agents that interact locally with one another and with the environment. The local interaction between agents will create a global behaviour.

Flock simulation is one of the SI-based algorithms which is used as a tool to simulate the interaction between natural flocks (e.g. bird flocks) [18]. A flock is a group of agents with similar properties that moves with respect to a steering force [20]. There are four kinds of flock steering forces: the alignment force, which allows each agent to align its heading with another agent within the same flock; the cohesion force, which keep together all agents in a flock; the separation force, which prevent crowding and collision within a flock; and the seek force, which steer a flock to a particular position [18]. Huang et al. [18] shows that flock simulation is suitable for simulating the movement of brush bristles throughout the canvas. The generated brush strokes are the flock of brush bristles that move together and leave a colour trail in the digital canvas.

3   Iterative Brush Stroke Path Extraction

3.1   Algorithm Overview

The primary assumption, as in most brush stroke extraction algorithms is that a brush path is characterized by a high level of homogeneity in colour values. Regions of similar colour are often processed using connected component labelling algorithms [15]. However, such methods are recursive in nature, and the stroke paths are obtained in a second pass where the shapes of the connected components are considered to extract medial lines.

The proposed algorithm is completely iterative in nature and simultaneously generates the stroke paths while identifying brush regions. A circular region of uniform colour (hereafter called a *blob*) is used as a brush template and its matching regions are iteratively computed, with the locus of the blob center forming a *brush path*. More formally, a blob of radius R with centre at pixel location P_0 is given by

$$B_R(P_0) = \{P \in I \mid \|P - P_0\| \leq R\} \tag{1}$$

where I represents the two-dimensional image space. The uniform colour constraint is expressed as a subset

$$S_R(P_0) = \left\{ P\hat{I}B_R(P_0) \mid \|v(P) - v(P_0)\| < DE \right\}$$
$$\#S_R(P_0) > 0.9\left(\#B_R(P_0)\right) \tag{2}$$

where $v(P)$ denotes the vector of colour components at pixel P, and ΔE is an error threshold used for colour comparison. In the second condition, the symbol # denotes the number of pixels in a set. This condition allows up to 10 % of pixels in a blob where the colour comparison has failed. We use a sequential left-to-right, top-to-bottom search using a set of radius number that will be fitted into the region until a matching region is found (Fig. 4).

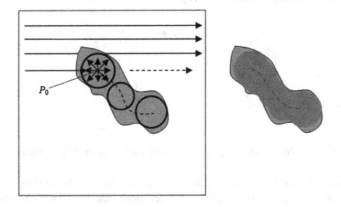

Fig. 4. Brush path extraction using a sequential region matching algorithm.

We then move the blob to its eight nearest neighbours, allowing a variation of the radius from R down to $R-10$, and select the direction that gives the maximum blob radius. This step that implements an adaptive size selection for the blob with a greedy path selection given by maximum radius forms the core of the algorithm. The centre points are marked as visited when a path is traced to avoid revisiting the same point in a subsequent iteration. A path is terminated when a blob of radius within the specified range is not found in the next step. At this point, all pixels within the matched blob regions are also marked as visited, and the search for the next blob location is resumed at the next unvisited pixel in the initial search path.

After scanning the whole image for blobs of radius R, we repeat the process by replacing R with a lower value $R \leftarrow R - \delta$ where δ is the step size, typically 2 units. Note that in each of the iteration, more and more pixels are marked as visited and therefore the processes for lower values of R take progressively lesser computational time.

3.2   Implementation Aspects

The L*a*b* (CIELAB) colour space is commonly used in image processing applications involving colour comparisons since it provides results that closely match the variations perceived by the human visual system [16]. In Eq. (3), we used a Euclidean distance measure with uniform weights and a difference threshold value of 3 (=ΔE):

$$\left(L^* - L^{*'}\right)^2 + \left(a^* - a^{*'}\right)^2 + \left(b^* - b^{*'}\right)^2 < 9 \tag{3}$$

The conversion from RGB to CIELAB is time consuming [17]. It is therefore preferable to convert all pixel values to CIELAB space at the beginning to avoid repeated computation of values for each blob region.

The adaptive scaling of blobs within a brush region can cause a smaller blob to be detected within a larger blob as shown in Fig. 5, resulting in "lumps" and thick lines in the path image. We removed such artefacts by using another set of flags for pixels within the detected blobs in the current brush region.

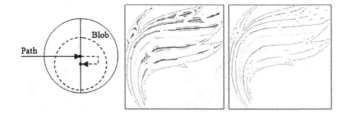

Fig. 5. Results showing the removal of thick lines in the path image.

The brush radii used along the 8 directions at the current position (Fig. 4) have integral values, and it is quite likely that we will get more than one direction with the same value for maximum radius. In such cases, we assigned a directional priority to indicate the preferred brush stroke path. For example, a left to right horizontal brush stroke can be given priority if that direction is among others that gave the maximum radius.

Such variation of reference colour value can often be seen across a brush stroke's region. The method shown in Fig. 4, uses the colour value at P_0 as the reference for the entire region. We could modify this approach by using the average of matched colour values within the current blob position as the reference for the next iteration, i.e., for the blobs searched along the 8 neighbouring directions. Such a variation of reference colour produces longer continuous paths instead of several broken segments.

When the blob radius reduces to a small value (for instance lower than 5 pixels), we require larger thresholds to accommodate extraneous pixels that do not have a colour that matches with the reference colour. For example, when the radius is less than 10, a 10 % threshold for extra pixels would give a value 0. Moreover at a small radius value, we get only a very coarse approximation of a circular shape. In general, regions corresponding to very thin brush strokes must be treated separately. Several such regions commonly appear in between larger segments of uniform colour. We use 3×3 and 5×5 square regions called "patches" to render small fragmented segments in the input image. The method of filling patches is distinctly different from that of blobs. Patches do not form a brush path; in other words, a patch does not have any "connected" patches. Secondly there is no reference colour for a patch. A patch is filled using a maximum voting method that determines the most commonly occurring colour value in that small segment. Thirdly, the size of the patch can vary based on the frequency of occurrence of a colour value in that region.

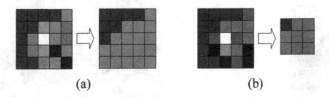

(a)   (b)

Fig. 6.   (a) The construction of a 5×5 patch and (b) a 3×3 patch.

In Fig. 6, a small region around a set of previously visited pixels is shown. The visited pixels are indicated in red colour, and the clear pixel is the current location (P_0) of the patch. In Fig. 6(a), there are 8 pixels of cyan colour in the 5×5 neighbourhood. The maximum frequency (or vote) of colours in the neighbourhood determines the size of the patch. Since we used 8 as the threshold, the colour cyan is used to fill the unvisited pixels in the 5×5 neighbourhood. The second example in Fig. 6(b) shows the colour green having a maximum vote of 5. In this case we use a 3×3 patch.

4   Results and Discussions

There are similarities in both approaches and outputs between algorithms for the extraction of edges and brush paths. An edge detection algorithm looks for gradients across regions while a brush path represents a direction of homogeneity in colour values. A brush path can be considered as a line (or multiple lines) between parallel edges of a region of nearly uniform colour. Both algorithms aim to provide outputs consisting of fewer fragmented lines, and this is achieved by pre-processing the image to get rid of high frequency components, and by careful selection of colour thresholds. In the experimental results shown below, we used Gaussian blur as a pre-processing step. Some aspects related to threshold selection have already been outlined in the previous section.

Three input images and the corresponding path images are given in Fig. 7(a) and (b). There are mainly two reasons for the discontinuities in the path images shown in Fig. 7(b). The iterative nature of the algorithm causes regions corresponding to large blob radius to be detected first, and smaller adjoining sections will be detected only in subsequent paths. Since a path always starts from the centre of a blob, there will be some gap between the two paths even if the two regions meet at a point. Secondly, there will be small intervening pixels with colour values outside the threshold required for matching with the reference colour, causing a path to terminate and a new path to begin after a few pixels. Increasing the threshold, on the other hand causes large blob regions to appear in the output. With a tighter threshold, only few blobs of large radius are seen (e.g. the stern of the boat in Fig. 7(a)). The sky image has large regions of nearly constant illumination and therefore results in longer paths with large blob radius. The effect of assigning directional priority to horizontal can be clearly seen in the path image, since otherwise paths could be in any arbitrary direction. The number of discontinuities along paths could be reduced using adaptive variations in the reference colour for the blobs along the path, as described in the previous section.

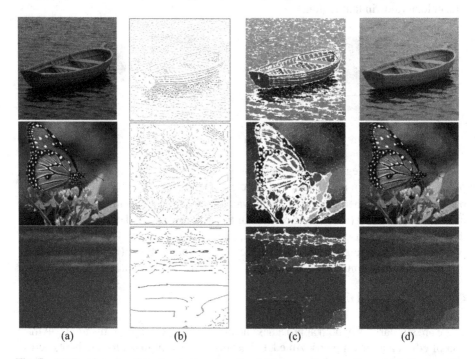

Fig. 7. (a) Test images, (b) extracted brush paths, (c) intermediate results and (d) final results showing the application of blob regions.

The extracted path can be used as guidance for brush bristles to move throughout the digital canvas in the flock simulation process. Experimental analysis shows that the medial lines obtained by our path extraction algorithm are more accurate than the ones obtained by applying connected component labelling and thinning as done by [18]. As a consequence, these lines will serve as better guide for the bristles.

A path image is actually a data structure that stores the radius and the blob's reference colour at each position. The painterly rendering of the input image is obtained by drawing the blobs at points where the radius value is non-zero. The intermediate results can be seen in Fig. 7(c). Small segments are then filled with patches before the final output (Fig. 7(d)) is obtained.

5   Conclusion and Future Work

This paper has presented an iterative method for identifying stroke regions and extracting brush stroke paths which closely match with the medial lines of the brush regions. Experimental analysis shows that these medial lines are more accurate than the ones obtained by applying connected component labelling and thinning. The iterative process of the extraction makes the processing faster than connected component labelling method which is done recursively. Several implementation aspects need to be considered in order to get proper brush directions so that stroke lines do not appear overly fragmented, and the painted output does not contain many undesirable artefacts. Most of these aspects have been discussed in detail and further explained using examples.

The immediate extension of this research is to use the generated medial lines for guiding the brush bristles in the flock simulation process. The brush bristles will be modeled using the four criteria of a flock as mentioned in Sect. 2. A flock centroid will be generated in the end of all lines. After that, a group of autonomous agents is normally distributed around the centroid. Every flock will have several control parameters such as: flock size, which specifies the number of agents inside the flock; energy; direction; alignment force; cohesion force; separation force; seek force; cohesion range; separation distance; gradient; and step size. The energy parameter specifies how many pixels it can travel across the canvas. A flock will stop moving if it runs out of energy. According to [18], a natural painting look can be achieved by modifying some flock parameters. For instance, the cohesion and separation force parameter will affect the brush stroke thickness. High cohesion force and low separation force will result to thin brush strokes, while high cohesion force and high separation force can lead to thick and broad brush strokes.

There are mainly two ways to look at the usefulness of the brush stroke extraction algorithm, as previously shown in Fig. 1. If the input image is a photograph, then the algorithm produces medial lines along which we can apply brush strokes to generate a painterly rendering of the input. If the input is a painting, we can use the medial lines to extract some information about the painting style in terms of density of strokes, length, curvature etc. For a painting, though, a brush region may not have the same level of homogeneity of colour as in a photograph. Streaks of thin lines formed by bristles and variations in colour are commonly seen within the same brush region in a painting. To account for these variations, we require higher level feature matching methods such as histograms or statistical descriptors. This would be the immediate and natural extension of the work presented in the paper.

Another possible extension of our research is to take the medial line properties as features for painting style characterization. Medial line features such as curvature and length along with the brushstrokes' geometrical and statistical features can characterize a particular style of a painter or an era from where a painting comes from [9, 11]. With the aid of machine learning techniques, the features can be used to differentiate paintings by an artist from another artist. It can also be used to classify artworks based on their art movement era.

References

1. Gooch, B., Gooch, A.: Non-photorealistic Rendering. AK Peters Ltd., Natick (2001)
2. Hertzmann, A.: A Survey of Stroke-based rendering. In: IEEE Computer Graphics and Application, vol. 4, pp. 70–81. IEEE Press, New York (2003)
3. Whitted, T.: Anti-aliased line drawing using brush extrusion. Assoc. Comput. Machi. Spec. Interest Group Comput. Graph. Interact. Techn. Comput. Graph. 17(3), 151–156 (1983). ACM
4. Strassman, S.: Hairy brush. Assoc. Comput. Machi. Spec. Interest Group Comput. Graph. Interact. Techn. Comput. Graph. 20(4), 225–232 (1986)
5. Pham, B.: Expressive brush strokes. Comput. Vis. Graph. Image Process. Graph. Models Image Process. 53(1), 1–6 (1991). Elsevier
6. Pudet, T.: Real time fitting of hand-sketched pressure brushstrokes. Comput. Graph. Forum 13(3), 205–220 (1994). Wiley Online Library
7. Hertzmann, A.: Painterly rendering with curved brush strokes of multiple sizes. In: 25th Annual Conference on Computer graphics and Interactive Techniques 1998, pp. 453–460. ACM (1998)
8. Huang, H.E., Lim, M.H., Chen, X., Ho, C.S.: Interactive GA flock brush for non-photorealistic rendering. In: Bui, L.T., Ong, Y.S., Hoai, N.X., Ishibuchi, H., Suganthan, P.N. (eds.) SEAL 2012. LNCS, vol. 7673, pp. 480–490. Springer, Heidelberg (2012)
9. Li, J., Yao, L., Hendriks, E., Wang, J.Z.: Rhythmic brushstrokes distinguish Van Gogh from his contemporaries: findings via automated brushstroke extraction. IEEE Trans. Pattern Anal. Mach. Intell. 34(6), 1159–1176 (2012). IEEE
10. Meer, P., Georgescu, B.: Edge detection with embedded confidence. IEEE Trans. Pattern Anal. Machine Intell. 23(12), 1351–1365 (2001). IEEE
11. Johnson Jr., C.R., Hendriks, E., Berezhnoy, I.J., Brevdo, E., Hughes, S.M., Daubechies, I., Li, J., Postma, E., Wang, J.Z.: Image processing for artist identification. IEEE Sig. Process. Mag. 25(4), 37–48 (2008). IEEE
12. Berezhnoy, I.E., Postma, E., Van Den Herik, H.J.: Automatic extraction of brushstroke orientation from paintings. Mach. Vis. Appl. 20(1), 1–9 (2009). Springer
13. Seo, S., Lee, H.: Pixel based stroke generation for painterly effect using maximum homogeneity neighbor filter. Multimedia Tools Appl. 74(10), 3317–3328 (2015). Springer
14. Gooch, B., Coombe, G., Shirley, P.: Artistic vision: painterly rendering using computer vision techniques. In: 2nd International Symposium on Non-photorealistic Animation and Rendering 2002, pp. 83-ff. ACM (2002)
15. Obaid, M., Mukundan, R., Bell, T.: Enhancement of moment based painterly rendering using connected components. In: International Conference on Computer Graphics, Imaging and Visualization 2006, Sydney, pp 378–383. IEEE (2006)
16. Reinhard, E., Khan, E.A., Akyuz, A.O., Johnson, G.: Colour Imaging: Fundamentals and Applications. AK Peters Ltd, Wellesley (2008)
17. Connolly, C., Fleiss, T.: A study of efficiency and accuracy in the transformation from RGB to CIELAB colour space. IEEE Trans. Image Process. 6(7), 1046–1048 (1997). IEEE
18. Huang, H. E., Ong, Y. S., Chen, X.: Autonomous flock brush for non-photorealistic rendering. In: IEEE Congress on Evolutionary Computation 2012, pp. 1–8. IEEE (2012)
19. Kennedy, J., Kennedy, J.F., Eberhart, R.C.: Swarm Intelligence. Morgan Kaufmann, San Francisco (2001)
20. Reynolds, C.W.: Flocks, herds and schools: a distributed behavioral model. ACM SIGGRAPH Comput. Graph. 4(21), 25–34 (1987). ACM

A Comparison Between Representations for Evolving Images

Alessandro Re[✉], Mauro Castelli, and Leonardo Vanneschi

NOVA IMS, Universidade Nova de Lisboa, 1070-312 Lisboa, Portugal
akirosspower@gmail.com

Abstract. Evolving images using genetic programming is a complex task and the representation of the solutions has an important impact on the performance of the system. In this paper, we present two novel representations for evolving images with genetic programming. Both these representations are based on the idea of recursively partitioning the space of an image. This idea distinguishes these representations from the ones that are currently most used in the literature. The first representation that we introduce partitions the space using rectangles, while the second one partitions using triangles. These two representations are compared to one of the most well known and frequently used expression-based representations, on five different test cases. The presented results clearly indicate the appropriateness of the proposed representations for evolving images. Also, we give experimental evidence of the fact that the proposed representations have a higher locality compared to the compared expression-based representation.

Keywords: Genetic programming (GP) · Image representation · Locality

1 Introduction

In Evolutionary Art, the problem of evolving images is still an open issue. Automatic and semi-automatic methods to generate interesting, but abstract[1] images have been studied for long time so far, and very interesting and appreciable results have been achieved. For instance [1] describes the Electric Sheep screensaver, a distributed system to evolve animated fractal structures, called *sheeps*, driven by user feedback. Another example is [2], where the author describes improved techniques for evolving images. Other examples can be found in [3,4]. On the other hand, the production of interesting images that represent precise, recognizable patterns, is still lacking. Recently, some authors have focused on this problem by adopting evolutionary techniques that combine a stochastic search algorithm, such as Genetic Programming (GP) [5,6] or Novelty Search [7], with pattern recognition models, such as Neural Networks, usually trained on a large set of pictures [8–11]. This approach is, in principle, very versatile, and has been

---

[1] In this context *abstract* is used to denote images that do not have to represent particular shapes or patterns, while *figurative* is used as its opposite.

© Springer International Publishing Switzerland 2016
C. Johnson et al. (Eds.): EvoMUSART 2016, LNCS 9596, pp. 163–185, 2016.
DOI: 10.1007/978-3-319-31008-4_12

used in specific domains, for instance to produce typefaces [12] and pictures [10]. However, it appears clear that one of the main concerns when developing these applications, is how to represent a solution, that is, in this case, an image. In recent studies [13], it has been shown that evolving a target image can be very hard. In fact, contrarily to the case of abstract images, in which the objective is to evolve "something" that has to satisfy some global criteria of "beauty", but does not have to necessarily have precise characteristics (like for instance particular figures or shapes), with figurative images, the goal is to reproduce a precise object or pattern (not necessarily representing a real object). In this case, as confirmed for instance in [13], the used representation has a major impact on the evolution of images. Therefore, using representations with different properties could make the search easier or harder. Given the hardness of this kind of application, it would be desirable to have representations that have high locality [14,15] (we say that a representation has *high locality* if small changes to genotype correspond to small changes to phenotype) and thus representations that can help us to define smooth fitness landscapes [16].

The objective of this paper is to present two new representations for the evolution of images, that can be particularly appropriate for the evolution of figurative images, due to their high locality. The main idea of these novel representations, that distinguishes them from the representations that are commonly used in the literature, is the concept of recursively partitioning the space of an image by means of geometric figures. More specifically, we introduce a representation in which the space is partitioned using rectangles (called Recursive Rectangles, or RR) and a representation in which the space is partitioned using triangles (called Recursive Triangles, or RT). To demonstrate the appropriateness of these two representations for generating figurative images, we compare them with one of the most used representations in the literature for the same task, i.e. an expression-based representation that we call EX. The experimental study is conduced on five different test cases, containing different shapes and shades. Results obtained using the RR, RT and EX representations are compared from several different viewpoints, including the qualitative resemblance between the evolved images and the target ones, the quantitative error between the evolved images and the target ones and locality.

The paper is structured as follows: in Sect. 2, we introduce the concepts of fitness landscapes and locality, and we discuss how they can be influenced by the choice of the representation. As a direct consequence, we also discuss how representations can influence the difficulty of a problem. In Sect. 3, we present a short overview of existing representations for the evolution of images, also presenting the EX representation and justifying the choice of comparing EX with RR and RT in our experimental study. In Sect. 4, we present our novel RR and RT representations. Section 5 contains our experimental study, in which RR and RT are compared to EX. After presenting and motivating the choice of the studied test cases, and describing the experimental setting in order to make the study completely replicable, we present and discuss the obtained experimental results. The three studied representations are compared qualitatively (in terms of the visual resemblance of the evolved images with the target) quantitatively (in terms of the error between the evolved images and the target) and also in terms of their locality. In the last part of Section 5, we also present and discuss the best images evolved using the different studied representations, highlighting the mutual differences and pointing out the various pros and cons. Finally, Sect. 6 concludes the work and suggests ideas for future work.

2   Representation, Fitness Landscapes and Locality

The representation of the solutions is a very important component of the configuration of an Evolutionary Algorithm (EA), and often the choice of the representation has a significant impact on the ability of the EA to solve a problem [17]. More specifically, it is a very diffused opinion in the literature that what makes a problem easy or hard to solve for an EA is, in large part, determined by the fitness landscape [16,18–21], and representations have an impact on the fitness landscape. A fitness landscape is a plot where the points in the horizontal direction represent the different individual genotypes in a search space, sorted according to the specific neighborhood induced by the variation operators used to explore the space itself (genetic operators for EAs) and the points in the vertical direction represent the fitness of each one of these individuals [5].

If we consider solution representations in EAs, it is clear that different representations can induce different neighborhoods (as studied in detail in [17]). As a simple example, one may think of the difference between the usual binary representation of numbers and the *Gray* binary representation [22] when searching for the maximum integer in a set given, as only operator to explore the space, the 1-bit-flip operator. If we think of fitness landscapes as three-dimensional maps, it is not difficult to convince oneself that changing the neighborhood corresponds to a reshuffling of the order of the solutions in the horizontal dimension, which may completely change the shape of the landscape itself.

Under this perspective, it is clear that changing the representation can radically change the fitness landscape, and thus the difficulty of a problem.

In Genetic Programming (GP) [5,6], the representation of solutions can be expressed by specifying two sets $\mathcal{F}$ and $\mathcal{T}$. $\mathcal{F}$ is the set of primitives, or functional symbols, and $\mathcal{T}$ is the set of terminal symbols used to encode programs. Thinking of the original representation of GP individuals as trees, introduced by Koza in his first book [6], functional symbols are the internal nodes of the trees, while terminals are the leaves. Fitness landscapes and problem difficulty in GP have been studied, for instance, in [23], where the importance of the representation, as a crucial choice to determine the ability of GP to solve a problem was widely demonstrated. It is nowadays clear that an inappropriate representation may make a problem harder to solve than an appropriate representation [17], as well as having significant importance over the execution time performance: a representation could be theoretically adequate (i.e. allowing to reach every solution by an iterative application of the operators), but practically unfit, because it will hinder the convergence to the desired optima.

In the last few years, in the GP community, noteworthy attention was dedicated to the concept of *locality*, as a measure of problem difficulty [14,15]. Broadly speaking, a configuration is said to be local (or to have high locality) if small differences in the syntactic structure of the individuals (small genotypic modifications) correspond to small differences in fitness (small phenotypic modifications). Typically, locality is measured by quantifying the difference in fitness between "neighbours", where two individuals x and y are said to be neighbours if x can be obtained by applying mutation to y, or viceversa (recent extensions to the concept of locality to also include crossover have been presented in [24,25]). Locality has a clear relationship with the shape of the fitness landscape; in particular, it is clear that a high locality corresponds to smooth fitness landscapes, while a low locality corresponds to rugged fitness landscapes.

In this work, we introduce two new representations for evolving images with GP. When designing these representations, as discussed in the continuation of this paper, we dedicated particular attention to locality. Thanks to their high locality, these representations should allow us to define smoother fitness landscapes, and thus improve the evolvability of GP.

3   Short Review of Existing Representations for Images

Although the possible ways of representing images can be numerous and diverse, they can be partitioned into two broad categories: resolution-dependent and resolution-independent representations. Typically, the former are raster images, while the latter are vectorial images.

Raster images are represented as a grid of colors, in which each cell is a pixel. Viewed from a certain distance, the eye does not perceive the grid, but perceives the image pattern. Raster images have a size, measured by number of pixels in each dimension (width and height). Naturally, the size of the image has an impact on the information content of the image, and therefore to its ability of representing a pattern. For instance, it is not possible to create an image of 2×3 pixels which can be unequivocally recognized as *The Mona Lisa*: a larger image is required for this task.

Vectorial images, on the other hand, are not tied to a specific image size, but are defined by discrete, mathematical entities: a line, for instance, could be expressed by providing an equation like $y = mx + q$; similarly, a circle could be represented by an equation like $x^2 + y^2 = r^2$. The entire image is a description of the elements that compose it, and how they interact to get the desired effects (e.g. how to fill non-simple polygons with one color). In CAD and vector graphic editing software, the primitive elements in images are usually lines, curves, polygons, gradients, colors, patterns and raster images. These elements can usually be grouped and combined in many different ways to yield a wide range of results, ranging from drawings to photorealistic paintings. Vectorial images normally require an additional step, called rendering, that translates the (usually) continuous, symbolic representation of the image into a discrete, raster image.

Vectorial representation is widely used in literature to generate figurative images. For example, the Scalable Vector Graphics (SVG) format can be used as representation for images [26,27], allowing to define images including lines, paths, fillings and shades, while vectorial representations (mostly limited to lines) have been used for typefaces evolution [12,28–30], or for evolving faces, where the evolution controls parameters of the output image [31]. It is noteworthy that these latter representations are not general purpose, but are specific for producing fonts and faces, while in the case of SVG evolution, the representation is general purpose. Another example of general purpose vectorial representation is the use of scalable geometrical primitives arbitrarily positioned in the space, such as the use of polygons or triangles to represent elements of the image, which can be arbitrarily colored or moved on the canvas [32].

The most common representation, introduced in [33], is called *expression-based representation*. This representation uses mathematical expressions employing common operators like for instance sum, subtraction, multiplication and division. Typically, an image is represented as a mathematical function in two variables, for example $f_1(x,y) = x^2 + y^2$ or $f_2(x,y) = \sin(x)\cos(y)$, where f_1 and f_2 are functions to be evaluated for each pixel in the image (see Fig. 1).

The expression-based representation is a widely diffused representation when evolving images: about 40% of the works cited in [34], regarding evolution of 2D images, uses expression based representation. Two possible causes of this diffusion are probably both historical and practical: not only this representation is one of firsts introduced in the field of Evolutionary Art, by the pioneering work of [33], but it also allows a large degree of customization, by altering the primitives used, still maintaining the tree-based representation widely diffused in the GP community.

Another representation which is conceptually similar to expression-based, but has a different structure and allows greater complexity, is by using Compositional Pattern Producing Networks (CPPNs) [35], which use mathematical entities as nodes of a network. CPPNs have been used in different works to represent image solutions [9,36].

One of the prominent critics to expression-based representations [37] is that the results usually look "too artificial". Indeed, the set of primitives $\mathcal{F}$ used to represent the individuals is often real-valued and the results are often smooth and soft-edged. On the other hand, many digital painting softwares include hard-edge tools and gradient tools, while typical primitive sets $\mathcal{F}$ for expression

Rendering of the function $f_1(x,y) = x^2 + y^2$

Rendering of the function $f_2(x,y) = \sin(x)\cos(y)$

Fig. 1. Rendering of the functions $f_1(x,y) = x^2 + y^2$ and $f_2(x,y) = \sin(x)\cos(y)$ on 8-bit grayscale images of size 400×200 pixels.

representations are mostly continuous, smooth primitives and few hard-edge functions (typically absolute value or conditionals). The problem could be solved by introducing new primitives for expression-based representations in the set $\mathcal{F}$, which are more similar to the tools used in digital painting softwares, for example splines and ellipses. This clearly modifies the nature of the mathematical objects: while the typical functional is a $\mathbb{R}^2 \rightarrow \mathbb{R}$ map, objects like splines, ellipses or circles are not functions at all. For instance, the circle $x^2 + y^2 = r^2$ is not a function of x and y. This problem could be solved by building distance functions for each of those objects. For example, a functional to draw a circle with center in (x_0, y_0) and radius r is:

$$f_c(x, y) = \begin{cases} 1 & \text{if } (x - x_0)^2 + (y - y_0)^2 \leq r^2 \\ 0 & \text{otherwise} \end{cases}.$$

We are not aware of publications that use specific geometrical shapes, defined in such a way, but techniques based on distance functions have been successfully used in the 3D Computer Graphics community, especially when applied to ray-marched shaders [38,39].

For domain specific applications, for example in the case of building type-faces [12], specific representations could be used. In that case, a resolution-independent representation has been used, but the description of the image is not (directly) based on mathematical symbols, but on a conceptual model in which straight lines could be placed over a discrete grid, where each grid point is a starting or ending point for one or more lines. In addition to starting and ending points, lines have a width that is variable.

In general, determining the most appropriate set of operators $\mathcal{F}$ for expression-based representations is a hard task and it is dependent on the application and on experience. The interested reader is referred to [40] for an in-depth analysis. In this work, we will use *expression-based* when referring to the general class of representations, and *EX* when referring to our specific variant.

4 The Proposed Representations

In this section we present the novel representations for images introduced in this paper. These representations aim at being general purpose, in the sense that they should be appropriate for representing any image, or at least a vast set of them. Also, we introduce these representations with the objective of overcoming a drawback of expression-based representations, namely evolvability. They are based on the concept of recursion. Recursion allows us to define representations that are very compact, in the sense that the number of primitives used is reduced, but with a relatively high level of flexibility concerning the possible visual outcomes. The fundamental idea behind these novel representations is that the functional primitives have the objective of partitioning the space into sub-spaces, each of which can be filled with a color or can be further partitioned, recursively. Specifically, the first one of these representations depicts an image by partitioning the space into rectangles, and thus we call it *Recursive Rectangle* (RR) representation. The second one partitions the space into triangles, and thus we call it *Recursive Triangle* (RT) representation.

The idea of partitioning the space recalls the concept of fractals, which have been used by some authors for evolutionary art, see for example [41,42]. Also, a similar approach has been used in [43] to define low-complexity art, in which legal circles are used as a basic unit to support the creation of artistic images.

The partition could be made with other geometries as well, not only triangles and rectangles. Partitioning the space means that it can be divided in smaller non-overlapping areas, leaving no empty spaces (this operation is also called tessellation or tiling). In general, there is no restriction on the shape used to partition the space, as long as it is possible to perform the partition in a finite number of recursive steps. On the other hand, it is convenient to use shapes that translate to few, low-complexity primitives for evolving the image using GP. For example, triangles and rectangles are well suited to partition the space, because any triangle can be divided easily into 2, 3 or 4 triangles (self-similar or not), and rectangles can be divided easily into 2 or 4 rectangles; on the other hand, circles or ellipses cannot be used to partition the space in a finite number of recursive steps.

The RR representation uses a set of primitives $\mathcal{F}$ composed by two functionals and a set of terminals $\mathcal{T}$ composed by a set of symbols whose number varies depending on the bit-depth used, i.e. the number of bits that are used to represent (the color of) a pixel. So, for instance, when images are bitmaps with just two colors, we use a set $\mathcal{T}$ that contains only two terminal symbols. When more colors are used, a larger set of constants can be used to represent each color. The set $\mathcal{F}$ contains two primitives: *vertical split* (VSplit) and *horizontal split* (HSplit). They partition the space according to their orientation. In this way, a space of size (w, h) will be partitioned vertically into two spaces of size $(w/2, h)$ or horizontally partitioned into two spaces of size $(w, h/2)$. An example of this partitioning is shown in Fig. 2. It is worth pointing out that any raster image can

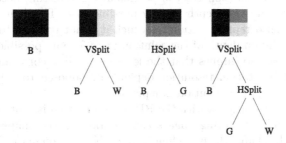

Fig. 2. Various examples of images and their corresponding encoding using the proposed RR representation. In this example, B stands for black, W for white and G for gray. For every subtree, the space managed by that subtree is: (1) partitioned horizontally or vertically, according to the root of the subtree being HSplit or VSplit, or (2) filled with a color if the subtree is composed by just one terminal symbol.

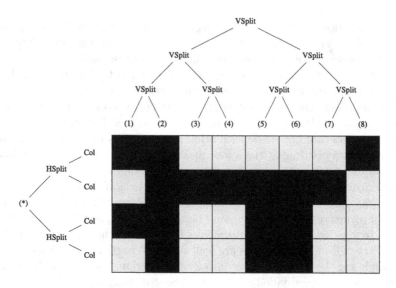

Fig. 3. An example of how a binary raster image of size $(8, 4)$ can be represented by RR. The subtree on the left of the image, rooted by symbol (∗), has to replace all the subtrees at level three of the tree represented on top.

be ultimately represented by RR, if no tree height restrictions are imposed. In fact, assuming that the image dimensions are power of two (i.e. $(w, h) = (2^n, 2^m)$ for some m, n), then it is possible to partition the image in exactly $w \times h$ pixels by applying each functional, VSplit and HSplit, $w - 1$ and $h - 1$ times respectively, yielding a tree of height $\log_2(w) + \log_2(h)$. For example, given a binary image of size $(8, 4)$, it is possible to build a tree where every terminal symbol is mapped to exactly one pixel in the image. Such tree is, for example, the one shown in Fig. 3.

The ability to control the granularity of the image, by imposing a maximum height for the solution tree, is particularly interesting because it allows to derive a maximum tree depth for any possible case. This is in contrast with the expression-based representation, in which it is not possible to determine in advance the maximum depth of the solution trees for any possible target.

Naturally, this also implies that the level of detail is constrained, therefore making our representation resolution-dependent, opposed to expression-based representations that are resolution-independent.

One of the main concerns with the RR representation is that we expect it to have problems in representing images that contain curved shapes. In fact, the more curved is the shape, the more fine grained the solution has to be to precisely follow it. This implies deeper trees, with more nodes and terminals, making it hard to precisely optimize the areas containing curves. To partially overcome this limitation, the RT representation partitions the space using triangles, allowing to get skewed lines easily. In particular, in the RT representation the set $\mathcal{F}$ contains just one primitive, that we call *triangle split*, or TSplit; this primitive partitions

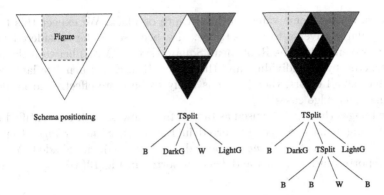

Fig. 4. Various examples of images and their corresponding encoding using the proposed RT representation. It is worth pointing out, as shown in the leftmost picture, that the main triangle, that defines the whole space partitioned by the nested recursive calls of the TSplit primitive, *includes* the image (which is, of course, rectangular). The rectangular area of the image is reported using dashed lines. In the central and rightmost pictures, B stands for black, W for white, DarkG for dark gray and LightG for light gray.

the space in 4 similar triangles, top (or bottom, depending on orientation), left, center and right. An example of this partitioning is shown in Fig. 4.

Besides the extremely small number of functionals used by our novel representations RR and RT, a feature that provides a significant reduction of the dimension of the space of solutions, an interesting property of these representations is their *locality*: changing a subtree, a terminal or a node, affects only a part of the image (the one influenced by the modified subtree), without influencing all the other areas. In other words, small changes in a genotype correspond to small changes in the phenotype. Naturally, changing the root of the tree from e.g. a vertical split to an horizontal split, may have a large impact if measured as the number of changed pixels in the image, but the image does not increase or decrease its complexity. Therefore, also in this case, the change can be considered as a relatively small one. Experimental results comparing the locality of RR and RT to the one of the EX representation are reported in Sect. 5.8.

5   Experimental Study

5.1   Test Cases

To test the representations, we used target images that could be considered conceptually simple figures of shapes. Targets included both hard-edges and smooth, shaded patterns: three images out of five include shadings. We also included a photographic image, a classic image widely used in the image processing field: Lena. Recreating this image is undeniably harder than the others, not only for the complex patterns, but also due to the used colors, which is in the gray range, without having extreme values such as white or black. We expect this test to be particularly adverse to our settings, also because of the considered fitness function: we employed the Root Mean Square Error (RMSE) between the image reconstructed by an individual and the target. RMSE will penalize large errors more than small errors, therefore it is likely to be more effective in hard-edge cases than soft-edge cases.

The images that we have used as target, from now on, will be identified using the following names: *Shapes* (reported in Fig. 5(a)), *Shaded Shapes* (reported in Fig. 7(a)), *Yin and Yang* (reported in Fig. 9(a)), it Shaded Yin and Yang (reported in Fig. 11(a)) and *Lena* (reported in Fig. 13(a)).

5.2   Experimental Settings

To test the representations, we used tree-based standard GP [6], with subtree crossover (with probability 0.8), subtree mutation (with probability 0.1) and tournament selection of size 3. Population has been initialized using the ramped half-and-half method, with a maximum depth of 7.

Each run used a population of 500 individuals, that was allowed to evolve for 200 generations, a limit that we found appropriate to evolve interesting results, understanding evolution trajectories, but keep running times low. No elitism was used to increase diversity. We also performed experiments using a population size equal to 1000 individuals, but results were qualitatively very similar to the ones obtained using 500 individuals, therefore we will present the results obtained using 500 individuals. For all the target images, except Lena, we used only two terminals (black and white) for RR and RT, while for EX we used a 8-bit grayscale. This is likely hindering the results of RR and RT on shaded target images, but should work efficiently on the non-shaded targets. Therefore, we expect that the average behavior does not change significantly. For the Lena test case, on the other hand, we included 8 gray shades in the terminal symbols set. In fact, using only black and white would enormously disadvantage RR and RT with RMSE used as fitness, since the entire picture does not contain white or black. As discussed in Sect. 4, for RR we used two functionals: VSplit and HSplit, while for RT we used only the TSplit functional. For EX, on the other hand, we selected a combination of functionals among the most used [40]: if-then-else (ternary), sum, sub, mul, protected division, min, max, pow, sqrt, square, absolute, negative and sign.

5.3   Experimental Results Obtained on the Shapes Test Case

The results obtained on the Shapes test case are reported in Figs. 5 and 6. In particular, in Fig. 5 we report, for each one of the studied representations, the pixel-by-pixel average of the images reconstructed by the best individual in the population at termination, calculated over 30 independent runs (it is worth noticing that the images reconstructed by the best evolved individuals will be reported and discussed in Sect. 5.9). In Fig. 6 we report, for each one of the studied representations, the median best fitness for each generation, calculated on

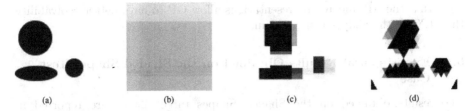

Fig. 5. Target image and images reconstructed using the different studied representations for the **Shapes** test case. Plot (a): target image. Plot (b): image reconstructed using the EX representation. Plot (c): image reconstructed using the RR representation. Plot (d): image reconstructed using the RT representation.

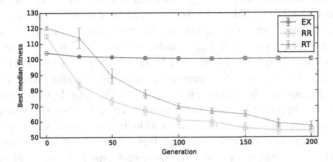

Fig. 6. Evolution of fitness, and its standard deviation, against generations for the three studied representations for the **Shapes** test case. For each generation, and for each one of the three representations, the reported fitness value is the median of the best fitness in the population at that generation, calculated over 30 independent runs. Standard deviations are represented using error bars.

the same 30 independent runs as Fig. 5. As Fig. 5(b) clearly shows, the EX representation is not appropriate for reconstructing the Shapes test image. On the other hand, the RT and RR representations seem to be more appropriate for this task, both producing images that clearly recall the Shapes target (the Shapes target is reported in Fig. 5(a)). Among the images reconstructed by the RT and RR representations (Fig. 5(c) and (d) respectively), it is hard to make a qualitative comparison, given that the result is highly subjective. Nevertheless, it is clearly possible to observe that RR produces more regular shapes, and less affected by noise (Fig. 5(c)), compared to RT. These results are qualitatively confirmed by the plots in Fig. 6, where we can see that the RR representation allows us to obtain images with a smaller pixel-by-pixel error compared to the other two representations. Looking at Fig. 6, we can also observe that in the final part of the evolution the results obtained using RR and RT are very similar. On the other hand, when EX is used, GP is practically unable to evolve: the best fitness in the population in the first part of the run is very similar to the best fitness at the end of the run, as we can see by the fact that the curve of EX in Fig. 6 is practically parallel to the horizontal axis. As a partial conclusion, we can state that the RR and RT representations allow GP to have better evolvability than EX for the Shapes test problem.

5.4   Experimental Results Obtained on the Shaded Shapes Test Case

The results obtained on the Shaded Shapes target image are reported in Figs. 7 and 8 where, as for the previous test case, Fig. 7 reports the pixel-by-pixel average images of the best individuals at termination obtained with each studied representation, and Fig. 8 reports the median best fitness vs. generations curves. Looking at Fig. 7, one may observe that none of the representations is able to reconstruct the Shaded Shapes target faithfully. However, it is worth pointing out that the objective of this work is not to obtain an effective reconstruction of the target images, but to compare the studied representations between each other. Under this perspective, Fig. 7 indicates that RT is the representation that allows us to reproduce the Shaded Shapes target in a qualitatively better way. More specifically, the EX representation is able to effectively reconstruct the shades that are in the target image, but totally unable to reconstruct the objects; the RR representation is able to reconstruct neither the shade, nor the objects; finally the RT representation is able to reproduce some shades even though only black and white are used as terminals, and it reconstructs shapes that recall the ones that are present in the target image. Figure 8 gives us some useful hints on the usefulness of the pixel-to-pixel RMSE between the target image and the reconstructed one as a fitness function. In fact, the representation that is able to obtain the image with the smallest error is clearly EX. This was possible because EX was able to faithfully reconstruct the background of the image, including the shade. Given that the background occupies the majority of the space in the image, EX results as the representation that was able to return the smallest error, even though it was unable to recreate any of the shapes that appear in the target. This clearly tells us that root mean square error is not a good fitness function for this kind of application. This fact was already know [44] and, for this reason, several alternative fitness functions have been

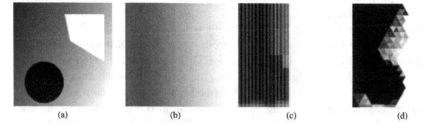

(a)   (b)   (c)   (d)

Fig. 7. Target image and images reconstructed using the different studied representations for the **Shaded Shapes** test case. Plot (a): target image. Plot (b): image reconstructed using the EX representation. Plot (c): image reconstructed using the RR representation. Plot (d): image reconstructed using the RT representation.

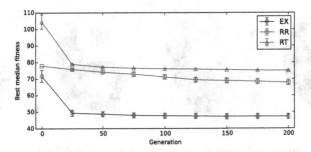

Fig. 8. Evolution of fitness, and its standard deviation, against generations for the three studied representations for the **Shaded Shapes** test case. For each generation, and for each one of the three representations, the reported fitness value is the median of the best fitness in the population at that generation, calculated over 30 independent runs.

proposed in the literature (for example [44], defines a fitness based on image complexity estimated by compression, while [45], uses edge-detection algorithms before comparing the images).

Also in this case, it is important to point out that the objective of this work is just a comparison between different representations under the same conditions, and not obtaining the best possible reconstruction of the target images. Being this the first step in our research track, we have decided to keep the study as simple as possible, and thus we have decided to use the widely used RMSE as fitness function, even though we are aware of its limitations.

5.5 Experimental Results Obtained on the Yin and Yang Test Case

The results obtained on the Yin and Yang target are reported in Figs. 9 and 10. Once again, as for the previous test cases, Fig. 9 reports the pixel-by-pixel average images of the best individuals at termination obtained using each one of the studied representations, and Fig. 10 reports the median best fitness vs. generations curves. Looking at Fig. 9, we can say that for this test case the advantage of the proposed RR and RT representations, compared to EX, is remarkable. In fact, EX generates an image that has nothing in common with the target (except maybe the presence of a light region in the upper part of the image and a dark region in the lower part). On the other hand, both RR and RT allow us to reconstruct the target rather faithfully, in the sense that in both cases the typical pattern of the Yin and Yang image is clearly perceivable in the reconstructed image. Also in this case, a comparison between the image reconstructed by RR and the one reconstructed by RT is not easy. However, we can remark that RT reconstructs in a more effective way than RR the little central black and white patterns (even though, as it was possible to expect, they have a broadly triangular shape, instead of circular as in the target). In fact, when RR is used, these features are both reproduced with an high degree of variation (the produced average images are mostly gray in those areas), while when RT is used,

Fig. 9. Target image and images reconstructed using the different studied representations for the **Yin and Yang** test case. Plot (a): target image. Plot (b): image reconstructed using the EX representation. Plot (c): image reconstructed using the RR representation. Plot (d): image reconstructed using the RT representation.

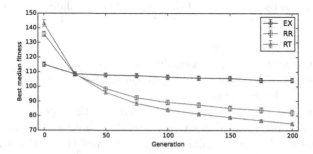

Fig. 10. Evolution of fitness, and its standard deviation, against generations for the three studied representations for the **Yin and Yang** test case. For each generation, and for each one of the three representations, the reported fitness value is the median of the best fitness in the population at that generation, calculated over 30 independent runs.

the upper one is completely black, and it appears in a completely white area, while the lower one is completely white, and it appears in a completely black area. This fact makes us slightly prefer the reconstruction made by the RT representation. This result is also confirmed by the curves reported in Fig. 10, where it is possible to see that RT is the representation that has allowed us to find the solutions with the smallest RMSE. It is interesting to point out that when the EX representation is used, GP is practically unable to evolve for this test problem (in the plot, the curve is practically parallel to the horizontal axis), while this is not the case with RR and RT.

5.6   Experimental Results Obtained on the Shaded Yin and Yang Test Case

The results obtained on the Shaded Yin and Yang Test Case (reported in Figs. 11 and 12) can be considered, in some senses, as consistent with both the ones obtained on Shaded Shapes (Sect. 5.4) and the ones obtained on Yin and Yang (Sect. 5.5). In fact, looking at the average of the best reconstructed

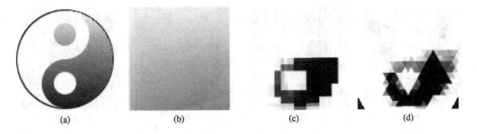

(a)   (b)   (c)   (d)

Fig. 11. Target image and images reconstructed using the different studied representations for the **Shaded Yin and Yang** test case. Plot (a): target image. Plot (b): image reconstructed using the EX representation. Plot (c): image reconstructed using the RR representation. Plot (d): image reconstructed using the RT representation.

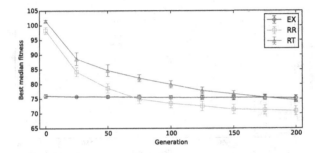

Fig. 12. Evolution of fitness, and its standard deviation, against generations for the three studied representations for the **Shaded Yin and Yang** test case. For each generation, and for each one of the three representations, the reported fitness value is the median of the best fitness in the population at that generation, calculated over 30 independent runs.

images reported in Fig. 11, we can see that, as for the other shaded target image previously discussed (Shaded Shapes), EX is able to reconstruct the shade, but it is totally unable to reconstruct the shape. As for the Yin and Yang test case, also for the Shaded Yin and Yang the image generated by EX does not have any element that makes us perceive what the target is. On the other hand, both RR and RT generate images that recall the typical Yin and Yang pattern. Also, as reported in Fig. 12, RR is the representation that allows us to find the solution with the smallest error, while, when EX is employed, GP is practically unable to evolve (once again, the curve of EX is practically parallel to the horizontal axis in the plot of Fig. 12).

5.7   Experimental Results Obtained on the Lena Test Case

As we can see from Figs. 13 and 14, the Lena target image is clearly the hardest one to reconstruct among all the test cases used so far. In fact, looking at the average reconstructed images reported in Fig. 13, we can see that none of the studied representations is able to generate an image that recalls us the target.

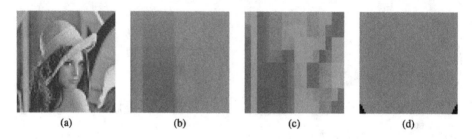

      (a)   (b)   (c)   (d)

Fig. 13. Target image and images reconstructed using the different studied representations for the **Lena** test case. Plot (a): target image. Plot (b): image reconstructed using the EX representation. Plot (c): image reconstructed using the RR representation. Plot (d): image reconstructed using the RT representation.

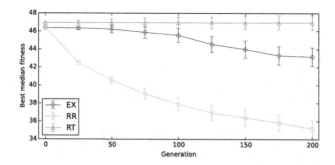

Fig. 14. Evolution of fitness against generations for the three studied representations for the **Lena** test case. For each generation, and for each one of the three representations, the reported fitness value is the median of the best fitness in the population at that generation, calculated over 30 independent runs.

Nevertheless, it is clearly possible to identify a difference between EX and RT, that both generated an almost constantly gray image with approximately the same color as the background of the target image, and RR in which some very fuzzy hints of some patterns that are present in the target can be grasped. One may for instance consider Lena's hair and hat, which are respectively darker and lighter then the background. In the corresponding positions, RR was able to generate areas that are, respectively, darker and lighter than the background. Also, observing the target image, it is possible to see a dark "diagonal line" on the bottom (right behind Lena's face), extending approximately until the upper right corner of the image (that is the border of the mirror in front of Lena). In the corresponding positions, RR was clearly able to generate a shape that is darker than the background. As reported in Fig. 14, RR is also the representation that has allowed us to find the solutions with the smallest RMSE.

5.8   Experimental Results on the Locality of the Representations

Locality is vastly studied in GP literature [14,15,24,25]. As already discussed, in broad terms, a representation is considered to be "local" if small variations in the genotype correspond to small variations in the phenotype. In other terms, a representation is local if the difference in fitness between and individual x and an individual obtained by applying mutation to x is "small". Locality has a clear relation with the shape of the fitness landscape. In fact, a low locality implies large differences in fitness between neighbors, which clearly causes ruggedness in the fitness landscape [14,15]. In several studies [14,15], locality has been considered as one of the causes of the effectiveness, or weakness, of a representation. In this work, we state that locality may be one of the reasons why RR and RT have allowed us to obtain better results than EX, as discussed above. Hence, the objective of this section is to compare the locality of the three studied representations. Several measures of locality have been defined so far [14,15,24,25], as well as several ways of calculating it. In this work, we use a measure that is rather similar to the one used in [14]. In synthesis, we proceed as follows: we consider a sample of 1000 randomly generated individuals, $\mathcal{X} = \{T_1, T_2, ..., T_{1000}\}$. In our study, we generated 500 individuals using the grow algorithm, and the remaining 500 individuals using the full algorithm (we are currently evaluating more sophisticated sampling techniques, like for instance the ones defined in [23]). For each $i = 1, 2, ..., 1000$, we generate 50 "neighbors" of individual T_i: $\mathcal{N}_i = \{N_{(i,1)}, N_{(i,2)}, ..., N_{(i,50)}\}$, i.e. 50 individuals obtained by applying mutation to T_i. Then, for each $i = 1, 2, ..., 1000$, and for each $j = 1, 2, ..., 50$, we calculate the pixel-by-pixel distance between the image reconstructed by individual T_i and the one reconstructed by its neighbor $N_{(i,j)}$. In other words, if d is the pixel-by-pixel distance between images, for each $i = 1, 2, ..., 1000$, and for each $j = 1, 2, ..., 50$ we calculate $d_{(i,j)} = d(T_i, N_{(i,j)})$. As a next step, for each $i = 1, 2, ..., 1000$ we calculate the following average:

$$L_i = \frac{1}{50} \sum_{j=1}^{50} d_{(i,j)}$$

Finally, we calculate our indicator of locality $\mathcal{L}$, as follows:

$$\mathcal{L} = \frac{1}{1000} \sum_{i=1}^{1000} L_i$$

where the smaller $\mathcal{L}$ the higher will be the locality. In other words, for each individual T_i in our sample, we calculate the average of all the distances between T_i and all its neighbors, and we use the average of all these averages, calculated over all the individuals of the sample, as an indicator of locality. If we repeat this process for 30 independent times, we obtain 30 different values of $\mathcal{L}$, which are reported in Fig. 15 for each one of the studied distances, after having sorted them from the smallest to the largest. The pieces of information we obtain from this figure are: (1) our novel representations RR and RT have larger locality

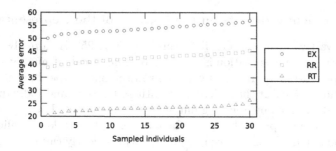

Fig. 15. The obtained results for the locality test performed on the three studied representations, sorted from the smallest to the largest. The values represent the average amount of change (error) of the fitness due to mutation to randomly generated individuals. As it can be seen, the error introduced by mutation in the EX representation is higher than the error of our novel representations, meaning that EX is the least local representation and RT being the most local representation.

Table 1. Average and standard deviation of the 30 points reported in Fig. 15 for each one of the studied representations.

Representation	EX	RR	RT
Average	53.7995	42.4417	23.2342
Standard deviation	1.7098	1.6974	1.1489

than EX; (2) among RR and RT, the representation that has the higher locality is clearly RT.

For each one of the studied representations, we have also calculated the average of the 30 points reported in Fig. 15, along with their standard deviations. The numeric results are reported in Table 1. Table 1 gives a clear and synthetic vision of the fact that RT has a much higher locality than RR, and RR has a much higher locality than EX. To analyze the statistical significance of these results, a set of statistical tests has been performed on the mean errors. Being the data normally distributed according to D'Agostino and Pearson's test [46,47], Z-test has been used to test if the underlying distributions of the data produced with RR and RT have the same mean of the distribution of EX. A value of $\alpha = 0.01$ (with a Bonferroni correction) has been used. The p-values obtained are zero (less then 10^{-30}), both when comparing RR to EX and when RT is compared to EX. These results undoubtedly confirm that our novel representations are more local than the existing EX representation.

5.9   Best Evolved Individuals

As a last step of our experimental study, in this section we report the images reconstructed by the best individuals using the three studied representations. These images are shown in Fig. 16, where the results for each test problem are reported on a separate column, and the results obtained using the different

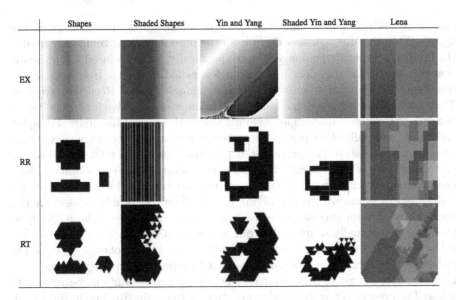

Fig. 16. For each one of the studied test problems, and for each one of the studied representations, we report the image reconstructed by best individual evolved by GP, over the 30 performed runs. More in particular, for each test problem (reported on columns), the topmost image was obtained using the EX representation, the one in the middle using the RR representation and the lowest using the RT representation.

representations are reported by row (top row for EX, central row for RR and bottom row for RT).

These images give a clear visual rendering about the suitability of the novel representation in generating images that, compared to the existing EX representation, more closely resemble the targets. Moreover, RT is the representation that has allowed us to obtain the results that we consider qualitatively better. Our interpretation of these results is that RT has a much higher locality, and thus allows GP to have a better evolvability. This should pave the way to the use of the RT representation in figurative arts.

6   Conclusions and Future Work

The importance of representations in Evolutionary Computing is widely recognized in the literature. In particular, the representation is one of the fundamental components that allow us to define a fitness landscape, and fitness landscapes are known to have an important impact on the difficulty of a problem. In recent studies, it has been shown that evolving a target image can be a very hard task. In this paper, we try to induce different fitness landscape by introducing two novel representations for images. These representations were designed with the explicit objective of improving locality, and thus inducing a smoother fitness landscape, compared to other existing representations. They are based on the idea of recursively partitioning the space of an image by means of geometric figures. This idea distinguishes them from the representations commonly used in the literature for evolving images. More specifically, we introduced a representation that partitions the space into rectangles, called Recursive Rectangle, and a representation that partitions the space into triangles, called Recursive Triangle. They have the advantage of having a small number of primitive functions (two in the case of Recursive Rectangle, and only one in the case of Recursive Triangle), and can be generalized, by partitioning the space by means of other geometric figures. In order to asses the appropriateness of these two novel representations, we performed a set of experiments. In these experiments, these representations have been compared with an expression-based representation, which is the one typically used in literature when evolving images. The test problems used for this comparison contain precise patterns, hard edges and shades. The majority of them are conceptually simple, while one of them is a very hard test case because it has a very detailed content and because of the vast gamma of colors it uses. The objective of our experimental study was not to reproduce as faithfully as possible the target images, but to compare the studied representations between each other. The presented results clearly indicate that Recursive Rectangle and Recursive Triangle are able to outperform the expression based representation, both from the qualitative viewpoint of the visual "resemblance" between the evolved images and the target and from the quantitative viewpoint of the root mean square error between the evolved images and the target. Further experiments have clearly demonstrated that Recursive Rectangle and Recursive Triangle are more local compared to the expression-based representation, and we hypothesize that this is one of the reasons why the two novel proposed representations have allowed us to obtain better results.

The work presented in this paper represents a first and preliminary step in a long and stimulating research track, whose final objective is to develop a system able to "create" novel, and possibly artistically valiant, images representing concepts, instead of merely reproducing existing ones. Thus, the final objective of our work is much harder than the one of this paper. Nevertheless, we believe that, before undertaking such an ambitious challenge, it was necessary to develop new representations that could outperform the existing ones at least for the easier and more basic problem of evolving target images. In particular, we believe that the locality of the used representations will be a crucial point of our future work. The work presented in this paper should pave the way to the use of Recursive Rectangle and Recursive Triangle (or their extensions using more sophisticated sets of geometric figures), for more ambitious tasks. Our current work is focused on improvements of Recursive Rectangle and Recursive Triangle. For instance, we are currently working on methods to adapt the presented representations to shaded images, by introducing shaded terminal symbols. Also, we are extending the presented work by considering different fitness functions, which have been shown in the literature to be more appropriate than the root mean square error. For instance, we are currently studying fitness functions based on data compression, like the one presented in [44]. Furthermore, we are developing methods to improve diversity inside populations of evolving images. Last but not least, we are currently designing possible extensions of the proposed representations, that can contribute to make them more general purpose, and thus improve their ability to generate images of different nature like, for instance, images based on line drawings. Such extensions will be tested and compared against expression-based representations as well as other interesting and general purpose representations such as CPPNs, using richer sets of operators, primitives and terminals.

References

1. Draves, S.: The electric sheep screen-saver: a case study in aesthetic evolution. In: Rothlauf, F., et al. (eds.) EvoWorkshops 2005. LNCS, vol. 3449, pp. 458–467. Springer, Heidelberg (2005)
2. Hart, D.A.: Toward greater artistic control for interactive evolution of images and animation. In: Giacobini, M. (ed.) EvoWorkshops 2007. LNCS, vol. 4448, pp. 527–536. Springer, Heidelberg (2007)
3. Romero, J.J., Machado, P. (eds.): The Art of Artificial Evolution: A Handbook on Evolutionary Art and Music. Natural Computing Series. Springer, Heidelberg (2008)
4. World, L.: Aesthetic selection: the evolutionary art of Steven Rooke [about the cover]. Comput. Graph. Appl. **16**(1), 4 (1996)
5. Langdon, W.B., Poli, R.: Foundations of Genetic Programming. Springer, Heidelberg (2002)
6. Koza, J.R.: Genetic Programming. The MIT Press, Cambridge (1992)
7. Lehman, J., Stanley, K.O.: Abandoning objectives: evolution through the search for novelty alone. Evol. Comput. **19**(2), 189–223 (2011)
8. Baluja, S., Pomerleau, D., Jochem, T.: Towards automated artificial evolution for computer-generated images. Connect. Sci. **6**(2–3), 325–354 (1994)
9. Nguyen, A., Yosinski, J., Clune, J.: Innovation engines: automated creativity and improved stochastic optimization via deep learning. In: Proceedings of the Genetic and Evolutionary Computation Conference (2015)
10. Correia, J., Machado, P., Romero, J., Carballal, A.: Evolving figurative images using expression-based evolutionary art. In: Proceedings of the Fourth International Conference on Computational Creativity, p. 24 (2013)
11. Machado, P., Correia, J., Romero, J.: Expression-based evolution of faces. In: Machado, P., Romero, J., Carballal, A. (eds.) EvoMUSART 2012. LNCS, vol. 7247, pp. 187–198. Springer, Heidelberg (2012)
12. Martins, T., Correia, J., Costa, E., Machado, P.: Evotype: evolutionary type design. In: Johnson, C., Carballal, A., Correia, J. (eds.) EvoMUSART 2015. LNCS, vol. 9027, pp. 136–147. Springer, Heidelberg (2015)
13. Woolley, B.G., Stanley, K.O.: On the deleterious effects of a priori objectives on evolution and representation. In: Proceedings of the 13th Annual Conference on Genetic and Evolutionary Computation, pp. 957–964. ACM (2011)
14. Galván-López, E., McDermott, J., O'Neill, M., Brabazon, A.: Defining locality as a problem difficulty measure in genetic programming. Genet. Program Evolvable Mach. **12**(4), 365–401 (2011)

15. Galvan, E., Trujillo, L., McDermott, J., Kattan, A.: Locality in continuous fitness-valued cases and genetic programming difficulty. In: Schütze, O., Coello Coello, C.A., Tantar, A.-A., Tantar, E., Bouvry, P., Del Moral, P., Legrand, P. (eds.) EVOLVE - A Bridge Between Probability, Set Oriented Numerics, and Evolutionary Computation II. AISC, vol. 175, pp. 41–56. Springer, Heidelberg (2012)

16. Stadler, P.F.: Fitness landscapes. In: Biological Evolution and Statistical Physics, pp. 183–204. Springer, Heidelberg (2002)

17. Rothlauf, F.: Design of representations and search operators. In: Kacprzyk, J., Pedrycz, W. (eds.) Springer Handbook of Computational Intelligence, pp. 1061–1083. Springer, Heidelberg (2015)

18. Deb, K., Goldberg, D.E.: Analyzing deception in trap functions. In: Whitley, D. (ed.) FOGA-2, pp. 93–108. Morgan Kaufmann, San Francisco (1993)

19. Horn, J., Goldberg, D.E.: Genetic algorithm difficulty and the modality of the fitness landscapes. In: Whitley, D., Vose, M. (eds.) FOGA-3, pp. 243–269. Morgan Kaufmann, San Francisco (1995)

20. Mitchell, M., Forrest, S., Holland, J.: The royal road for genetic algorithms: fitness landscapes and GA performance. In: Varela, F.J., Bourgine, P. (eds.) Toward a Practice of Autonomous Systems: Proceedings of the First European Conference on Artificial Life, pp. 245–254. MIT Press, Cambridge (1996)

21. Forrest, S., Mitchell, M.: What makes a problem hard for a genetic algorithm? Some anomalous results and their explanation. Mach. Learn. 13, 285–319 (1993)

22. Chakraborty, U.K., Janikow, C.Z.: An analysis of gray versus binary encoding in genetic search. Inf. Sci. 156(3–4), 253–269 (2003)

23. Vanneschi, L.: Theory and Practice for Efficient Genetic Programming. Ph.D. thesis, Faculty of Sciences, University of Lausanne, Switzerland (2004)

24. Uy, N.Q., Hoai, N.X., O'Neill, M., McKay, B.: The role of syntactic and semantic locality of crossover in genetic programming. In: Schaefer, R., Cotta, C., Kołodziej, J., Rudolph, G. (eds.) PPSN XI. LNCS, vol. 6239, pp. 533–542. Springer, Heidelberg (2010)

25. Uy, N.Q., Hoai, N.X., O'Neill, M., McKay, R., Phong, D.N.: On the roles of semantic locality of crossover in genetic programming. Inf. Sci. 235, 195–213 (2013). Data-based Control. Decision, Scheduling and Fault Diagnostics

26. den Heijer, E., Eiben, A.E.: Evolving pop art using scalable vector graphics. In: Machado, P., Romero, J., Carballal, A. (eds.) EvoMUSART 2012. LNCS, vol. 7247, pp. 48–59. Springer, Heidelberg (2012)

27. den Heijer, E., Eiben, A.E.: Evolving art with scalable vector graphics. In: Proceedings of the 13th Annual Conference on Genetic and Evolutionary Computation, pp. 427–434. ACM (2011)

28. Baker, E., Seltzer, M.: Evolving line drawings. In: Proceedings of the Fifth International Conference on Genetic Algorithms, pp. 91–100. Morgan Kaufmann Publishers (1994)

29. Unemi, T., Soda, M.: An IEC-based support system for font design. In: IEEE International Conference on Systems, Man and Cybernetics, vol. 1, pp. 968–973. IEEE (2003)

30. Schmitz, M.: genoTyp, an experiment about genetic typography. In: Proceedings of Generative Art 2004 (2004)

31. Pagliarini, L., Parisi, D.: Face-it project. In: Proceedings of XV Italian Congress on Experimental Psychology, pp. 38–41 (1996)

32. Alsing, R.: Genetic programming: Evolution of Mona Lisa (2008). http://rogeralsing.com/2008/12/07/genetic-programming-evolution-of-mona-lisa/

33. Sims, K.: Artificial evolution for computer graphics, vol. 25(4), pp. 319–328. ACM (1991)
34. Lewis, M.: Evolutionary visual art and design. In: Romero, J., Machado, P. (eds.) The Art of Artificial Evolution: A Handbook on Evolutionary Art and Music. Natural Computing Series, pp. 3–37. Springer, Heidelberg (2008)
35. Stanley, K.O.: Compositional pattern producing networks: a novel abstraction of development. Genet. Program Evolvable Mach. 8(2), 131–162 (2007)
36. Secretan, J., Beato, N., D'Ambrosio, D.B., Rodriguez, A., Campbell, A., Folsom-Kovarik, J.T., Stanley, K.O.: Picbreeder: a case study in collaborative evolutionary exploration of design space. Evol. Comput. 19(3), 373–403 (2011)
37. McCormack, J.: Open problems in evolutionary music and art. In: Rothlauf, F., et al. (eds.) EvoWorkshops 2005. LNCS, vol. 3449, pp. 428–436. Springer, Heidelberg (2005)
38. Jiarathanakul, P.: Ray marching distance fields in real-time on webgl. Technical report, Citeseer
39. Quilez, I.: Modeling with distance functions (2008). http://iquilezles.org/www/articles/distfunctions/distfunctions.htm
40. Greenfield, G.R., et al.: Mathematical building blocks for evolving expressions. In: Bridges: Mathematical Connections in Art, Music, and Science, pp. 61–70. Tarquin Publications (2000)
41. Ventrella, J.J.: Evolving the mandelbrot set to imitate figurative art. In: Hingston, P.F., Barone, L.C., Michalewicz, Z. (eds.) Design by Evolution: Advances in Evolutionary Design. Natural Computing Series, pp. 145–167. Springer, Heidelberg (2008)
42. Lutton, E., Cayla, E., Chapuis, J.: $ArtiE - fract$: the artist's viewpoint. In: Cagnoni, S., et al. (eds.) EvoIASP 2003, EvoWorkshops 2003, EvoSTIM 2003, EvoROB/EvoRobot 2003, EvoCOP 2003, EvoBIO 2003, and EvoMUSART 2003. LNCS, vol. 2611, pp. 510–521. Springer, Heidelberg (2003)
43. Schmidhuber, J.: Low-complexity art. Leonardo, 97–103 (1997). JSTOR
44. Machado, P., Cardoso, A.: All the truth about NEvAr. Appl. Intell. 16(2), 101–118 (2002)
45. Di Gesu, V., Starovoitov, V.: Distance-based functions for image comparison. Pattern Recogn. Lett. 20(2), 207–214 (1999)
46. D'Agostino, R.B.: An omnibus test of normality for moderate and large size samples. Biometrika 58(2), 341–348 (1971)
47. D'Agostino, R.B., Pearson, E.S.: Tests for departure from normality. empirical results for the distributions of b2 and b1. Biometrika 60(3), 613–622 (1973)

Evolving L-Systems with Musical Notes

Ana Rodrigues$^{(\boxtimes)}$, Ernesto Costa, Amílcar Cardoso, Penousal Machado,
and Tiago Cruz

CISUC, Deparment of Informatics Engineering, University of Coimbra,
Coimbra, Portugal
{anatr,ernesto,amilcar,machado,tjcruz}@dei.uc.pt

Abstract. Over the years researchers have been interested in devising computational approaches for music and image generation. Some of the approaches rely on generative rewriting systems like L-systems. More recently, some authors questioned the interplay of music and images, that is, how we can use one type to drive the other. In this paper we present a new method for the algorithmic generations of images that are the result of a visual interpretation of an L-system. The main novelty of our approach is based on the fact that the L-system itself is the result of an evolutionary process guided by musical elements. Musical notes are decomposed into elements – pitch, duration and volume in the current implementation – and each of them is mapped into corresponding parameters of the L-system – currently line length, width, color and turning angle. We describe the architecture of our system, based on a multi-agent simulation environment, and show the results of some experiments that provide support to our approach.

Keywords: Evolutionary environment · Generative music · Interactive genetic algorithms · L-systems · Sound visualization

1 Introduction

It is a truism to say that we live in a world of increasing complexity. This is not because the natural world (physical, biological) has changed, but rather because our comprehension of that same world is deeper. On the other hand, as human beings, our artificial constructions and expressions, be them economic, social, cultural or artistic, are also becoming more complex. With the appearance of the computers, the pace of complexification of our world is increasing, and we face today new fascinating challenges. Computers also gave us a new kind of tool for apprehending and harnessing our world (either natural or artificial) through the lens of computational models and simulations. In particular, it is possible to use the computer as an instrument to interactively create, explore and share new constructs and the ideas behind them.

Music is a complex art, universally appreciated, whose study has been an object of interest over the years. Since the ancient days, humans have developed a natural tendency to translate non-visual objects, like music, into visual codes,

C. Johnson et al. (Eds.): EvoMUSART 2016, LNCS 9596, pp. 186–201, 2016.
DOI: 10.1007/978-3-319-31008-4_13

i.e., images, as a way to better understand those artistic creations. More recently, some authors have tried to translate images into sounds using a wide variety of techniques. Although there is still a lot of work to be done in the field of cross-modal relationships between sound and image [1–7], the achievements made so far in the devising of audio-visual mappings show that this approach may contribute to the understanding of music.

In this work we are interested in using computers to explore the links between visual and musical expressions. For that purpose we develop an evolutionary audiovisual environment that engages the user in an exploratory process of discovery.

Many researchers have been interested in devising computational approaches for music and image generation. Some of these approaches rely on generative rewriting systems like L-systems. More recently, some authors questioned the interplay of music and images, that is, how can we use one type to drive the other. Although we can find many examples of L-systems used to algorithmic music generation [1,5,7–9], it is not so common to find generation of L-systems with music. Even less common is to find attempts to have it working in both ways.

We present a new method for the algorithmic generation of images that are the result of a standard visual interpretation of an L-system.

A novel aspect of our approach is the fact that the L-system itself is the result of an evolutionary process guided by musical elements. Musical notes are decomposed into elements – pitch, duration and volume in the current implementation – and each of them is mapped into corresponding parameters of the L-system – currently line length, width, color and turning angle.

The evolution of the visual expressions and music sequences occurs in a multi-agent system scenario, where the L-systems are the agents inhabiting a world populated with MIDI (Musical Instrument Digital Interface) musical notes, which are resources that these agents seek to absorb. The sequence of notes collected by the agent, while walking randomly in the environment, constitutes a melody that is visually expressed based on the current interpretation of the agent's L-system. We use an Evolutionary Algorithm (EA) to evolve the sequence of notes and, as a consequence, the corresponding L-system. The EA is interactive, so the user is responsible for assigning a fitness value to the melodies [10].

The visual expression provided by the L-system aims to offer visual clues of specific musical characteristics of the sequence, to facilitate comparisons between individuals. We rely on tools such as Max/Msp to interpret the melodies generated and Processing to build the mechanisms behind the interactive tool and the respective visual representations.

Even if the main focus is to investigate the L-systems growth with musical notes, we also try to balance art and technology in a meaningful way. More specifically, we explore ways of modeling the growth and development of visual constructs with music, as well as musical content selection based only on the visualization of the constructs.

Moreover, we are also interested in understanding in which ways this visual representation of music will allow the user to associate certain kinds of visual patterns to specific characteristics of the corresponding music (e.g., its pleasantness). The experiments made and the results achieved so far provide support to our approach.

The remainder of the paper is organized as follows. In Sect. 2, we present some background concepts needed to understand our proposal. In Sect. 3, we describe some work related with the problem of music and image relationship. In Sect. 4 we specify the system's architecture and development, which includes describing the audiovisual mappings and the evolutionary algorithm we use. We continue in Sect. 5 with the presentation of the results. Lastly, in Sect. 6, we present our main conclusions, achieved goals and future improvements.

2   Background

In this section we briefly refer to the main concepts involved in the three basic elements of our approach: L-systems, evolutionary algorithms and music.

2.1   L-Systems

Lindenmayer Systems, or L-systems, are parallel rewriting systems operating on strings of symbols, first proposed by Aristid Lindenmayer to study the development processes that occur in multicellular organisms like plants [6]. Formally, an L-system is a tuple $G = (V, \omega, P)$, where V is a non-empty set of symbols, ω is a special sequence of symbols of V called axiom, and P is a set of productions, also called rewrite rules, of the form $LHS \rightarrow RHS$. LHS is a non-empty sequence of symbols of V and the RHS a sequence of symbols of V. An example of L-systems is:

$$G = (\{F, [,], +\}, F, \{F \rightarrow F[F][+F]\})$$

As a generative system, a L-system works by, starting with the axiom, iteratively rewriting in parallel all symbols that appear in a string using the production rules. Using the previous example of L-system we obtain the following rewritings:

$$F \xrightarrow{1} F[F][+F] \xrightarrow{2} F[F][+F][F[F][+F]][+F[F][+F]] \xrightarrow{3} \dots$$

After **n** rewritings we say we obtain a string of level **n**. The axiom is considered the string of level 0.

In order to be useful as a model, the symbols that appear in the string must be interpreted as elements of a certain structure. A classical interpretation, that we will use here, is the turtle interpretation, first proposed by Prusinkiewicz [11]. The symbols of a string are commands for a turtle that is moving in a 2D world. The state of the turtle is defined by two attributes: position (x, y) and orientation α. The commands change these attributes, eventually with side-effects (e.g., drawing a line). In Table 1 we show this interpretation.

Table 1. Turtle interpretation of an L-system

Symbol	Interpretation
F	Go forward and draw a line
f	Go forward without drawing
+	Turn counter-clockwise
-	Turn counter-clockwise
[	Push turtle's state
]	Pop turtle's state

Fig. 1. Example of a visual interpretation of the string at level 5.

Using this interpretation the visual expression of the string of level 5, of the given L-system is presented in Fig. 1. Notice that the user has to define two parameters: the step size of the forward movement and the turn angle.

Over the years L-systems were extended and their domains of application, both theoretical and practical, was broadened [1,5,7–9]. Some L-systems are context-free (the LHS of each production has at most one symbol), while others are context-sensitive (the production have the form $xAy \rightarrow xzy$, with $A \in V$ and $x, y, z \in V^+$. Some L-systems are said to be determinist (at most one production rule with the same left hand side) while others are stochastic. Some L-systems are linear while others, like the one above, are bracketed. The latter are used to generate tree-like structures. Yet some other L-systems are said to be parametric, i.e., when a parameter is attached to each symbol in a production rule whose application depends on the value of the parameter. Finally, some L-systems are called open, when they communicate with the environment and may change as a consequence of that communication [4].

2.2 Evolutionary Algorithms

Evolutionary Algorithms (EA) are stochastic search procedures inspired by the principle of natural selection and in genetics, that have been successfully applied in problems of optimization, design and learning [12]. They work by iteratively improving a set of candidate solutions, called individuals, each one initially generated at random positions. At each evolving step, or generation, a subset of promising solutions, called parents, is selected according to a fitness function for reproduction with stochastic variation operators, like mutation and crossover.

Mutation involves stochastic modifications of some components of one individual, while crossover creates new individuals by recombining two or more. The result of these manipulations is a new subset of candidate solutions, called offspring. From the parents and the offspring we select a new set of promising solutions, the survivors. The process is repeated until a certain termination criterion is met (e.g., a fixed number of generations). Usually the algorithm does not manipulate directly the solutions but, instead, a representation of those solutions, called the genotype. To determine the quality of the genotypes they must be mapped into a form that is amenable for the assessment by the fitness function, called phenotype.

2.3   Musical Concepts

Notes, or pitched sounds, are the basic elements of most music. Three of the most important features that characterise them are: pitch, duration and volume. *Pitch* is a perceptual property of sound that determines its highness or lowness. *Duration* refers to how long or short a musical note is. *Volume* relates to the loudness or intensity of a tone.

Most of the western music is *tonal*, i.e., melody and harmony are organised under a prominent tonal center, the tonality, which is the root of a major or minor scale.

When a central tone is not present in a music, it is said to be *atonal*.

Even though the concepts of harmony and progression do not apply in an atonal context, the quality of the sounding of two or more tones usually strongly depends on formal and harmonical musical contexts in which it occurs. This quality is usually classified as *consonance*. Consonance is a context-dependent concept that refers to two or more simultaneous sounds combined in a pleasant/agreeable unity of sound. On the other side, dissonance describes tension in sound, as if sounds or pitches did not blend together, and remain separate auditive entities [13]. Anyway, consonance is a relative concept: there are several levels of consonance/dissonance. Although consonance refers to simultaneous sounds, it may also be applied to two successive sounds due to the memorial retention of the first sound while the second is heard.

The difference between two pitches is called *interval*. Intervals may be harmonic (two simultaneous tones) or melodic (two successive tones). In tonal music theory, intervals are classified as perfect consonants (perfect unison and perfect 4th, 5th and 8th intervals), imperfect consonants (major and minor 3rd and 6th) and dissonants (all the others) [13].

Our system produces melodic sequences of notes in the C Major Scale. However, we do not constrain the system to produce tonal sequences or even consonant pairs of sounds.

3   Related Work

This section presents some of the most relevant references to the development of our work.

We can see in the following examples, that science and music have a long common history of mutual interactions. As Guéret et al. [14] say, music can lead to new perceptions in scientific realities and science can improve music performance, composition and understanding.

Music has a huge structural diversity and complexity. Algorithms that resemble and simulate natural phenomena are rich in geometric and dynamic properties. Computer models of L-systems are an example of such algorithms, and therefore can be helpful in automatic music composition [8,15].

Early work on L-systems and Music includes Keller and Měch et al. [4,16]. Many authors have described techniques to extract musical scores from Strings produced by L-systems [9,11,16]. One of the first works on the field of music generation and L-systems belongs to Prusinkiewicz [1]. He described a technique to extract music from a graphical interpretation of an L-system string. The length of each note was interpreted as the length of the branch, and the note pitch was interpreted as the vertical coordinate of the note. Graphical and musical interpretation were synchronized [11]. A survey on the evolution of L-systems in artistic domains includes McCormack work [2].

This mapping of sound parameters into something that usually is not considered audible data is called sonification. Many had an interest in exploring sonification in a way of understand scientific data the same way visualization is able to do [7].

There have been some efforts to create evolutionary systems for the automatic generation of music. A remarkable example is the EA proposed by Biles [17]. In this work he uses a GA to produce jazz solos over a given chord progression. In recent years several new approaches have emerged based not only on GA, but on other techniques such as Ant Colony [3,14,17–19].

4   System's Architecture

To explore the interplay between music and visual expressions by L-systems we construct a 2D world. In this section we describe this world, the entities that live and interact in it and evolve under the user guidance.

4.1   General Overview

In our world there are two types of entities: agents and notes. Notes have immutable attributes: their position and value. They do not die or evolve over time. Agents are entities with two components: (1) an L-system that drives its visual expression and (2) a sequence of notes that define the L-system's parameters at each level of rewriting (see Fig. 2). Agents move in the world by random walk, looking for notes that they copy internally and append to their sequence. These notes change over time through an Interactive Genetic Algorithm (IGA) [10].

The environment begins with a non-evolved individual (level 0) that wanders in the environment catching notes. In this case, its growth is determined by the musical notes that it catches, creating a sequence of notes. The first note caught makes it evolve to level 1, the second note to level 2 and so forth Fig. 3.[1]

---

[1] It is possible to catch some note that has been previously caught.

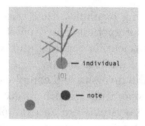

Fig. 2. Environment's elements. (Best viewed in color, see color figure online)

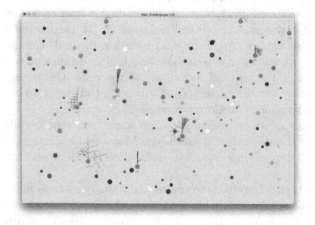

Fig. 3. Environment overview. (Best viewed in color, see color figure online)

A new individual can be generated from the current one through two different processes: mutation and crossover. A more detailed description of this can be found in Sect. 4.3.

When the user selects an individual there are two possible operations at the level of the interface: (i) Listen to the musical sequence of the individual in question note by note; (ii) Listen to the musical sequence of the individual in question in a simultaneous way (all notes together). Two other possibilities exist at the evolutionary level: (iii) Apply a mutation; (iv) Choose two different parents, and apply a crossover. (See Fig. 4)

4.2   Audiovisual Interpretation and Mappings

To have a qualitative criteria for auditory and visual mappings, we established a guide that formalises the relationship between these two domains:

1. Every auditory category admitted should have assigned a corresponding visual effect. We accomplish this by visualising the following parameters: (i) pitch, (ii) duration, (iii) volume, and (iv) notes interval.

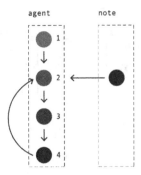

Fig. 4. System's architecture overview. (1) Agent is placed randomly in the environment; (2) Agent searches for a note and catches it; (3) Visual expression with an L-system of that note is made; (4) A GA can be applied.

2. As the work has a selection method based on an IGA, simplicity shall be maximised. When we have a high degree of complexity, the user often loses the ability to maintain sufficient visual control and perception over the results [20].

We divide the visual representation of music into two distinct parts: (i) the visual representation of the notes spread across the environment that individuals may catch, and (ii) the notes that the L-systems effectively catch. The first representation is static, because they are always a direct representation of the notes' parameters. The second one is dynamic, in the sense that different shapes are formed as new notes are caught.

Fig. 5. Graphic interpretation of the notes spread across the environment: (a) The note volume is represented by color saturation. The higher the note volume is, the more intense is the object's color. (b) Size represents the note duration. The higher the duration is, the bigger is the object's size.

The static notes in the environment are circles in levels of grey, with saturation representing volume and size representing note duration (see Fig. 5). Pitch is not represented. Position in the environment is random.

For the L-system visual representation, authors who have made similar attempts have chosen to associate the L-system's vertical coordinates to note pitch, and the distance between branches to note duration. However, we are

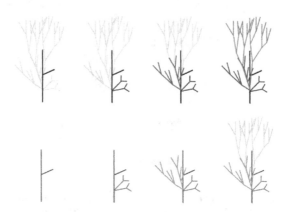

Fig. 6. Example of the L-systems growth process. (Best viewed in color) (Color figure online)

interested in comparing musical sequences in a qualitative way, considering the notion of consonance instead of absolute or relative pitch. Therefore, we had to adopt our own mappings between music and image to use in the L-system.

The L-systems presented in this work grow with the musical notes collected (see Fig. 6). Each note affects the L-system visual parameters at each level: (i) branch angle, (ii) branch length, (iii) branch weight, and (iv) color. Note duration maps into branch length, note volume into branch stroke (see Fig. 7), and consonance into branch color.

Every time a note is caught its pitch is compared to the previous note. From there, we calculate its consonance or dissonance. To the first note caught by an individual (level 1) is attributed a pitch color corresponding to its pitch height (see Fig. 8). If the sequence of notes is consonant then it is applied a tonality based on the color of the previous note caught. In case it is dissonant, a random color tonality is applied. Looking at Fig. 8 we can realize that consonance can be distinguished by its subtle change of color. On the contrary, a dissonant melody will produce changes of color and color tonalities with bigger steps.

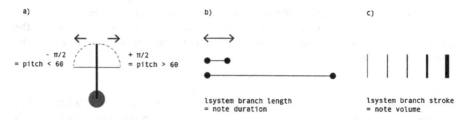

Fig. 7. Mapping process of the note's characteristics (a) pitch, (b) duration, (c) volume, into L-system graphic representation.

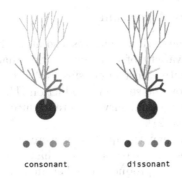

Fig. 8. Consonant and dissonant visual representations. (Best viewed in color) (Color figure online)

Furthermore, since there is no term of comparison to other notes when the L-system catches its first note, the color assigned corresponds to the pitch (see Fig. 9) of the caught tone. To the other notes color is assigned accordingly to the classification of consonant or dissonant depending on the note that has been previously caught.

Our environment is stochastic in the sense that agents walk randomly through the system. Furthermore the own process of note's modification implies a chance of being chose or not a note to apply these modifications. Stochastic systems can have different strings derived from the same string at any step, and they may produce a high diversity of sequences [7].

The number of possible outcomes for both sound and visual combinations is dependent on the number of possible values for the notes[2] pitch (127), duration (3800) and volume (82), in addition to the number of notes that we set up for each individual (4). Although we set the latter value to 4 in our experiments, it is not a limitation of our system.

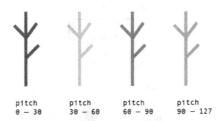

Fig. 9. Color association for pitch. Warm colors correspond to lower pitches, and cold colors to higher pitches. (Best viewed in color) (Color figure online)

---

[2] Each note parameters were interpreted as a MIDI note: (i) pitch range: 0 – 127 (ii) volume range: 20 – 102 (iii) duration range 200 – 4000 ms (iv) timbre – piano (0).

4.3   The Evolutionary Algorithm

Controlled evolution in the environment was a solution that we adopted to allow the creation of a large variety of complex entities that remain user directed and simple to interact with. Most organisms evolve by means of two primary processes: natural selection and sexual reproduction. The first determined which members of the population would survive to reproduce, and the second ensured mixing and recombination [21].

An IGA is used to assign the quality of a given candidate solution. The solutions favoured by the user have a better chance of prevailing in the gene pool, since they are able to reproduce in higher amount.

The musical sequence caught by an individual consists in its genotype, and its phenotype is composed of sound and image, i.e., L-system (see Fig. 10). The order of the genotype is defined by the order in which notes are caught.

Selection: Computationally, the measurement of the quality of a chromosome is achieved through a fitness function. In this work, this process is done interactively and is provided by a human observer. The use of an IGA, based in this case on the user visual and auditory perceptions, allows the user to direct evolutions in preferred directions. With this approach, the user gives real-time feedback. The expected output is a computer program that evolves in ways that resemble natural selection.

Offspring is born based on selected individuals, and to it a mutation process is applied. This replication of the preferred individual feeds up the probabilities of growing up more individuals that the user enjoys.

Reproduction: We apply both crossover and mutation in our system for evolution to progress with diversity. While crossover allows a global search on the

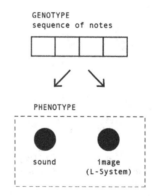

Fig. 10. The genotype (sequence of notes) is translated into sound and image (phenotype). Although sound has a direct mapping to MIDI notes, the image is interpreted with an L-system.

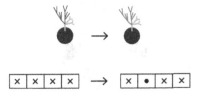

Fig. 11. Mutation example. One note of the original sequence of notes was chosen to be modified.

solutions space, mutation allows a local search. Each element has a chance (probability) of being mutated. Implementing these algorithms, we intend the evolution of L-systems with musical material through genetic transmission.

Offspring resulting from mutations or crossover are incrementally inserted into the current population and original chromosomes are kept. According to Sims [20], "Mutating and mating parameter sets allow a user to explore and combine samples in a given parameter space".

<u>Mutation</u>: Mutation takes a chromosome of an individual and randomly changes part of it [19]. It allows to change pitch, duration and volume in this case. Our mutation mechanism receives two parameters: the sequence of notes that will be modified and the probability of mutation of each note in the genotype. The probability of mutation will decide which note(s) collected by that individual will be modified. Each element in the sequence of notes caught by the individual has equal chance of being chosen (uniform probability). To each chosen note for mutation, the following parameters are changed randomly: pitch, duration and volume (see Fig. 11).

<u>Crossover</u>: Crossover allows the exchange of information between two or more chromosomes in the population [19]. This mixing allows creatures to evolve much more rapidly than they would if each offspring simply contained a copy of the genes of a single parent, modified occasionally by mutation [21]. In this case, it is possible to select only two parents which will give birth to two children.

We start by selecting random cut points on each parent, and then we give birth to the children based on these cut points (see Fig. 12). The resulting size of each child is variable since the cut points made in the parents are random.

4.4 Auxiliary Tools

To interpret sound we use Max/Msp. It is a graphical environment for creating computer music and multimedia works and uses a paradigm of graphical modules and connections. It reveals to be very helpful in sound interpretation and manipulation. For the grammatical construction and visual interpretation of L-systems we did rely on Processing [22]. Processing is a visual programming tool, suitable for designers and computer artists.

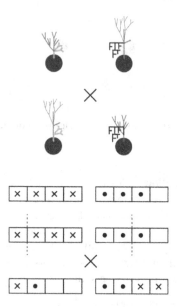

Fig. 12. 1 point crossover example. Two parents are crossed and give birth to two different children.

5   Experimental Results and Discussion

Music can be a very complex thing itself. When we add more complexity to it by using GAs and graphical interpretations of L-systems, if we are not careful, the perception and interaction of the system can easily get out of control. Given the experimental nature in this work, many of our decisions relied on simple concepts so that a full understanding of the system behavior would be possible.

According to Lourenço et al. [1], L-systems wouldn't be a perfect fit for this case because if the rendering techniques are too simple the resulting melody will probably end up with the same motif over and over again. Our solution to increase variability was to implement a generative solution and use some operators from GAs.

It is in fact far from trivial to conciliate both musical and pleasant aesthetic results with L-systems due to the small level of control of the structure. We have tried to solve this problem by providing the user the chance to interactively choose the survival chance of individuals. Although this system has been mostly guided through user interaction, we must question ourselves if it is possible to reach the same quality of results without user guidance.

Since all the parameters present on each L-system were translated into some kind of mapping, it had a direct impact on their developmental process. The resulting individuals revealed to have a lot of visual diversity and express well what we listen to as pleasant or not. Even though we work with simple musical inputs, a big variety of images and melodies (audiovisual experience) was produced as well (see Fig. 13).

Fig. 13. Example of multiple individuals generated by the system. (Best viewed in color) (Color figure online)

In sum, this audiovisual environment provides the user with a visual representation to a sequence of notes and visual pattern association to the musical contents which can be identified as pleasant or not pleasant. This also means that the user does not have to listen to every individual present in the environment to understand its musical relevance.

A demonstration video can be found at the following link: https://goo.gl/mrbhYa.

6 Conclusion and Further Work

The key idea that makes our approach different from others studies is the concern of mapping sound into image and image into sound. More specifically, our L-systems develop and grow according to the musical notes that were collected by them. At the same time, visual patterns aim to reflect musical melodies built in this process.

For the system evolution and subjective evaluation we have implemented a GA inspired in EC. Stronger individuals had higher probability to survive and reproduce while the weaker did disappear from the environment much faster. The use of an IGA allowed the user to work interactively and in novel ways, meaning that he/she would not be able to reach some results if the implemented computer generative system did not exist. Overall, the system hereby presented is an audiovisual environment that offers a rich and enticing user experience. This provides the user a clear and intuitive visual experience, which is something that we need to have into account since it is a system that is guided by the user.

In future work we would like to make an attempt implementing an ant colony behaviour for notes collection in the environment. It would also be important to investigate a more sophisticated process of music composition, including some rules of harmonisation and chord progression as well as the possibility to introduce more than one timbre in the system. Departing from a tonal system we could have then a set of musical rules that could lead to a fitness evaluation with more values. We have interest as well in exploring these audiovisual mappings at a perceptual level, i.e., using emotions provoked by music as a basis to guide the visual representations. Other future explorations could include L-system with a major diversity of expression or even the use of other biological organisms.

Acknowledgments. This research is partially funded by the project ConCreTe. Project ConCreTe acknowledges financial support of the Future and Emerging Technologies (FET) programme with the Seventh Framework Programme for Research of the European Commission, under FET grant number 611733.

References

1. Lourenço, B.F., Ralha, J.C., Brandao, M.C.: L-systems, scores, and evolutionary techniques. In: Proceedings of the SMC 2009–6th Sound and Music Computing Conference, pp. 113–118 (2009)
2. McCormack, J.: Aesthetic evolution of L-systems revisited. In: Raidl, G.R., Cagnoni, S., Branke, J., Corne, D.W., Drechsler, R., Jin, Y., Johnson, C.G., Machado, P., Marchiori, E., Rothlauf, F., Smith, G.D., Squillero, G. (eds.) EvoWorkshops 2004. LNCS, vol. 3005, pp. 477–488. Springer, Heidelberg (2004)

3. Moroni, A., Manzolli, J., Von Zuben, F., Gudwin, R.: Vox populi: an interactive evolutionary system for algorithmic music composition. Leonardo Music J. **10**, 49–54 (2000)
4. Měch, R., Prusinkiewicz, P.: Visual models of plants interacting with their environment. In: Proceedings of the 23rd Annual Conference on Computer Graphics and Interactive Techniques. SIGGRAPH 1996, PP. 397–410. ACM, New York (1996)
5. Pestana, P.: Lindenmayer systems and the harmony of fractals. Chaotic Model. Simul. **1**(1), 91–99 (2012)
6. Prusinkiewicz, P., Lindenmayer, A.: The algorithmic beauty of plants. Springer, New York (1990)
7. Soddell, F., Soddell, J.: Microbes and music. In: Mizoguchi, R., Slaney, J.K. (eds.) PRICAI 2000. LNCS, vol. 1886. Springer, Heidelberg (2000)
8. Kaliakatsos-Papakostas, M.A., Floros, A., Vrahatis, M.N.: Intelligent generation of rhythmic sequences using finite l-systems. In: Eighth International Conference on Intelligent Information Hiding and Multimedia Signal Processing (IIH-MSP), pp. 424–427. IEEE (2012)
9. Nelson, G.L.: Real time transformation of musical material with fractal algorithms. Comput. Math. Appl. **32**(1), 109–116 (1996)
10. Sims, K.: Interactive evolution of dynamical systems. In: Toward a practice of autonomous systems, Proceedings of the First European Conference on Artificial Life, pp. 171–178 (1992)
11. Prusinkiewicz, P.: Score Generation with L-Systems. MPublishing, University of Michigan Library, Ann Arbor (1986)
12. Eiben, A., Smith, J.E.: Introduction to Evolutionary Computation. Natural Computing Series, 2nd edn. Springer, Heidelberg (2015)
13. van Dillen, O.: Consonance and dissonance (2014). http://www.oscarvandillen. com/outline_of_basic_music_theory/consonance_and_dissonance/. Accessed 01 November 2015
14. Guéret, C., Monmarché, N., Slimane, M.: Ants can play music. In: Dorigo, M., Birattari, M., Blum, C., Gambardella, L.M., Mondada, F., Stützle, T. (eds.) ANTS 2004. LNCS, vol. 3172, pp. 310–317. Springer, Heidelberg (2004)
15. Manousakis, S.: Musical l-systems. Koninklijk Conservatorium. (master thesis), The Hague (2006)
16. Keller, R.M., Morrison, D.R.: A grammatical approach to automatic improvisation. In: Proceedings, Fourth Sound and Music Conference, Lefkada, Greece, July 2007, Mostof the soloists at Birdland had to wait for Parker's next record in order to and out what to play next. What will they do now (2007)
17. Biles, J.: Genjam: A genetic algorithm for generating jazz solos. In: Proceedings of the International Computer Music Conference, International Computer Music Association, pp. 131–131 (1994)
18. Todd, P., Werner, G.: Frankensteinian methods for evolutionary music composition. In: Griffith, N., Todd, P.M. (eds.) Musical Networks: Parallel Distributed Perception and Performance, pp. 313–339. MIT Press/Bradford Books, Cambridge (1999)
19. Wiggins, G., Papadopoulos, G., Phon-Amnuaisuk, S., Tuson, A.: Evolutionary methods for musical composition. Dai Research Paper (1998)
20. Sims, K.: Artificial evolution for computer graphics. Comput. Graphics **25**(4), 319–328 (1991)
21. Holland, J.H.: Genetic algorithms. Sci. Am. **267**(1), 66–72 (1992)
22. Shiffman, D.: Learning Processing: A Beginner's Guide to Programming Images, Animation, and Interaction. Morgan Kaufmann, Amsterdam (2009)

MetaCompose: A Compositional Evolutionary Music Composer

Marco Scirea[1(✉)], Julian Togelius[2], Peter Eklund[1], and Sebastian Risi[1]

[1] IT University of Copenhagen, Copenhagen, Denmark
msci@itu.dk
[2] New York University, New York, USA
http://www.itu.dk/msci

Abstract. This paper describes a compositional, extensible framework for music composition and a user study to systematically evaluate its core components. These components include a graph traversal-based chord sequence generator, a search-based melody generator and a pattern-based accompaniment generator. An important contribution of this paper is the melody generator which uses a novel evolutionary technique combining FI-2POP and multi-objective optimization. A participant-based evaluation overwhelmingly confirms that all current components of the framework combine effectively to create harmonious, pleasant and interesting compositions.

Keywords: Evolutionary computing · Genetic algorithm · Music generator

1 Introduction

Computer music generation is an active research field encompassing a wide range of approaches [1]. There are many reasons for wanting to build a computer system that can competently generate music. One is that music has the power to evoke moods and emotions – even music generated algorithmically [2]. In some cases the main purpose of a music generation algorithm is to evoke a particular mood. This is true for music generators that form part of highly interactive systems, such as those supporting computer games. In such systems a common goal of music generation is to elicit a particular mood that dynamically suits the current state of the game play. Music generation for computer games can be seen in the content of an experience-driven procedural content generation framework (*EDPCG*) [3], where the game adaptation mechanism generates music with a particular mood or affect expression in response to player actions.

While the medium-term goal of our system (described in Sect. 3) focuses on this kind of affect expression, this paper describes work on the core aspects of music generation, without expressly considering affective impact.

In games, unlike in traditional sequential media such as novels or movies, events unfold in response to player input. Therefore, a music composer for an interactive environment needs to create music that is dynamic while also being

© Springer International Publishing Switzerland 2016
C. Johnson et al. (Eds.): EvoMUSART 2016, LNCS 9596, pp. 202–217, 2016.
DOI: 10.1007/978-3-319-31008-4_14

non-repetitive. This applies to a wide range of games but not all of them; for example rhythm games make use of semi-static music around which the gameplay is constructed. The central research question in this paper is how to create music that is dynamic and responsive in real-time, maintaining fundamental musical characteristics such as harmony, rhythm, etc. We describe a component-based framework for (i) the generation of a musical abstraction and (ii) real-time music creation through improvisation. Apart from the description of the method the main research question of this paper is **validating the music generation algorithm: do all the components of the system add to the music generated?** To that end we present and discuss the results of a participant-based evaluation study.

2 Background

2.1 Procedurally Generated Music

Procedural generation of music is a field that has received much attention over the last decade. The approaches taken are diverse, they range from creating simple sound effects, to avoiding repetition when playing human-authored music, to creating more complex harmonic and melodic structures [4]. A variety of different approaches to procedural music generation have been developed, which can be divided into: *transformational* and *generative* algorithms [5]. MetaCompose falls in the latter category as it creates music without having any predefined snippets to modify or recombine.

Similar work to ours can be found in the system described by Robertson [6], which focuses on expressing fear. There are some parallels with this work, as the representation of the musical data through an abstraction (in their case the CHARM representation [7]), yet we claim our system has a higher affective expressiveness since it aims to express multiple moods via music. There are many examples of using evolutionary approaches to generating music, two example are the work of Loughran *et al.* [8] and Dahlstedt's evolution of piano pieces [9], many more can be found in the *Evolutionary Computer Music* book [10].

Other examples of real-time music generation can be found in some patents: a system that allows the user to play a solo over some generative music [11] and a system that can create complete concerts in real time [12]. An interesting parallel between the second system and ours is the incorporation of measures of "distance" between music snippets to reduce repetition. Still, both these approaches present no explicit affective expression techniques.

The work of Livingstone [13] in trying to define a dynamic music environment where music tracks adjust in real-time to the emotions of the game character (or game state). While this work is interesting, it is still limited (in our opinion), by the usage of predefined music tracks for affective expression. Finally *Mezzo*[14], a system designed by Daniel Brown that composes neo-Romantic game soundtracks in real time, creates music that adapts to emotional states of the character, mainly through the manipulation of *leitmotifs*.

2.2   Multi-Objective Optimization

Multi-Objective Optimization (MOO) is defined as the process of simultaneously optimizing multiple objective functions. In most multi-objective optimization problems, there is no single solution that simultaneously optimizes every objective. In this case, the objective functions are said to be partially conflicting, and there exists a (possibly infinite) number of Pareto optimal solutions. A solution is called nondominated, Pareto optimal, Pareto efficient or noninferior, if none of the objective functions can be improved in value without degrading some of the other objective values. Therefore, a practical approach to multi-objective optimization is to investigate a set of solutions (the best-known Pareto set) that represent the Pareto optimal set as much as possible [15]. Many Multi-Objective Optimization approaches using Genetic Algorithms (GAs) have been developed. The literature on the topic is vast; Coello lists more than 2000 references on this topic in his website[1].

Our approach builds on the successful and popular NSGA-II algorithm [16]. The objective of the NSGA-II algorithm is to improve the adaptive fit of a population of candidate solutions to a Pareto front constrained by a set of objective functions. The population is sorted into a hierarchy of sub-populations based on the ordering of Pareto dominance. Similarity between members of each subgroup is evaluated on the Pareto front, and the resulting groups and similarity measures are used to promote a diverse front of non-dominated solutions.

2.3   Feasible/Infeasible 2-Population Genetic Algorithm

Many search/optimization problems have not only one or several numerical objectives, but also a number of constraints – binary conditions that need to be satisfied for a solution to be valid. The approach we adopted for melody generation contains such strong rules, that are described in detail in Sect. 5.2. A number of constraint handling techniques have been developed to deal with such cases within evolutionary algorithms. FI-2POP [17] is a constrained evolutionary algorithm that keeps two populations evolving in parallel, where feasible solutions are selected and bred to improve their objective function values while infeasible solutions are selected and bred to reduce their constraint violations. Each generation, individuals are tested to see if they violate the constraints; if so they are moved to the 'Infeasible' population, otherwise they are moved to the 'Feasible' one. An interesting feature of this algorithm is that the infeasible population influences, and sometimes dominates, the genetic material of the optimal solution. Since the infeasible population is not evaluated by the objective function it cannot get stuck in a sub-optimal solution, but it is free to explore boundary regions, where the optimum is most likely to be found.

---

[1] http://www.cs.cinvestav.mx/~constraint/papers/.

3   MetaCompose

The presented system is composed of three main components: (i) composition generator, (ii) real-time affective music composer and (iii) an archive of previous compositions. The modular nature of the system allows components to be easily swapped for other components or augmented with further components. The archive (iii) maintains a database of all the previous compositions connected to the respective levels/scenes of the game. The archive allows persistence of compositions for later reuse, but also allows us to compute a measure of novelty for future compositions compared with what has already been heard. This database could also be extended to connect compositions to specific characters, events, game levels, etc. The real-time affective music composer is the component that transforms a composition in the final score according to a specific mood or affective state. The system is designed to be able to react to game events, such events depending on the effect desired, examples of responses to such events include a simple change in the affective state, a variation of the current composition or an entirely new composition.

4   Non-dominated Sorting Feasible-Infeasible 2 Populations

Usually, when dealing with constrained optimization problems, the solution adopted is the introduction of penalty functions to act as constraints. This approach favors the feasible solutions to the infeasible ones, potentially removing infeasible individuals that might lead to an optimal solution, and getting the solutions stuck at a local optimum. There have been many examples of a constrained multi-objective optimization algorithms [18–21].

Presented here is a combination of FI-2POP and NSGA-II dubbed Non-dominated Sorting Feasible-Infeasible 2 Populations (NSFI-2POP), uniting the benefits of keeping an infeasible population, which is free to explore the solution space without being dominated by the objective fitness function(s), and finding the Pareto optimal solution for multiple objectives. The algorithm takes the structure of FI-2POP, but the objective function of the feasible function is substituted with the NSGA-II algorithm. In Sect. 5.2 is described an application of this approach to the evolution of melodies.

5   Composition Generation

Composition in this paper is a *chord sequence*, a *melody* and an *accompaniment*. It is worth noting that the *accompaniment* is only an abstraction and not a complete score of a possible accompaniment, which is described in detail in Sect. 5.3 below. The main reason for the deconstruction of compositions is that we want a general structure (an abstraction) that makes music recognizable and gives it identity. Generating abstractions, which themselves lack some

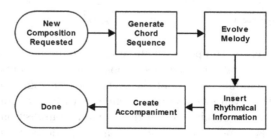

Fig. 1. Steps for generating a *composition*.

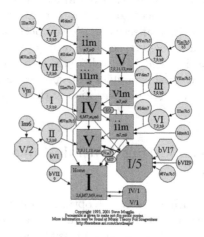

Fig. 2. Common chord progression map for major scales, created by Steve Mugglin [22].

information that one would include in a classically composed piece of music, e.g. tempo, dynamics, etc., allows the system to modify the music played in real-time depending on the affective state the game-play wishes to convey. The generation of compositions is a process with multiple steps: (i) create a chord sequence, (ii) evolve a melody fitting this chord sequence, and (iii) create an accompaniment for the melody/chord sequence combination (see Fig. 1).

5.1   Chord Sequence Generation

The method for generating a chord sequence works as follows: we use a directed graph of common chord sequences and preform random walks on this graph (see Fig. 2) starting from a given chord. As can be seen from the Fig. 2, the graph does not use a specific key, but rather 'degrees': in music theory, a degree (or scale degree) is the name given to a particular note of a scale to specify its position relative to the 'tonic' (the main note of the scale). The tonic is considered to be the first degree of the scale, from which each octave is assumed to begin. The degrees in Fig. 2 are expressed in Roman numerals and, when talking about chords, the numeral in upper-case letters symbolizes a major chord, while lower-case letters (usually followed by an m) express a minor chord. Other possible variations on the chord are generally expressed with numbers and other symbols, which we don't list for the sake of brevity. So, if we consider the D major scale, the $Dmajor$ chord would correspond to a I degree, while a $iiim$ degree would be a $F\sharp minor$. Various parameters of this sequence can be specified, such as sequence length, first element, last element, chord to which the last element can resolve properly (e.g., if we specify that we want the last chord to be able to resolve in the V degree, the last element might be a IV or a iim degree).

An interesting aspect of this graph is that it also shows common resolutions to chords outside of the current key, which provide a simple way of dealing with key changes. Each chord can be interpreted as a different degree depending on which key is considered, so if we want a key change we can simply: (i) find out which degree the last chord in the sequence will be in the new key and (ii) follow the graph to return to the new key. This produces harmonious key changes which do not sound abrupt.

5.2   Melody Generation

Melodies are generated with an evolutionary approach. We define a number of features to both include and avoid in melodies based on classical music composition guidelines and personal musical practice. These features are divided into constraints and objective functions. Accordingly, we use a Feasible/Infeasible two-population method (FI-$2POP$ [17]) with multi-objective optimization [23] for the Feasible population. Given a chord sequence, a variable number of notes is generated for each chord, which will evolve without duration information. Once the sequence of notes is created, we generate the duration of the notes randomly.

Genome Representation. The evolutionary genome consists of a number of values (the number of notes to be generated) that can express the notes belonging to two octaves of a generic key (i.e. 0–13). Here, we do not introduce notes that do not belong to the key, effectively making the context in which the melodies are generated strictly diatonic. Alterations will appear in later stages, in the real-time affective music composer module, when introducing variations of the composition to express affective states or chord variations.

Constraints. We have three constraints: a melody should: (i) *not have leaps between notes bigger than a fifth*, (ii) *contain at least a minimum amount of leaps of a second* (50 % in the current implementation) and (iii) *each note pitch should be different than the preceding one*.

$$\text{Feasibleness} = -\sum_{i=0}^{n-1}(\text{Second}(i, i+1) + \text{BigLeap}(i, i+1) + \text{Repeat}(i, i+1))$$

where n is the genome length   (1)

The three functions comprising Eq. 1 are all boolean functions that return either 1 or 0 depending if the two notes at the specified indexes of the genome satisfy the constraint or not. As can be seen, this function returns a number that ranges from (potentially) $-\infty$ to 0, where reaching the 0 score determines that the individual satisfies all the constraints and, consequently, can be moved from the unfeasible population to the feasible one.

On the constraints in Eq. 1, leaps larger than a fifth do appear in music but they are avoided here, as experience suggest they are too hard on the ear of the listener. Namely, if the listener is not properly prepared, leaps larger than a fifth can easily break the flow of the melody. We also specify a minimum number of intervals of a second (the smallest interval possible considering a diatonic context such as this one, see the *Genome Representation* section) because if the melody has too many larger leaps it feels more unstructured, and not something that we would normally hear or expect a voice to sing. Finally the constraint on repetition of notes is justified by the fact that these will be introduced in the real-time interpretation of the abstraction.

Fitness Functions. Three objectives build to compose the fitness functions: the melody should (i) *approach and follow big leaps (larger than a second) in a counter step-wise motion (explained below)* (Eq. 2), (ii) *where the melody presents big leaps the leap notes should belong to the underlying chord* (Eq. 3) and finally (iii) *the first note played on a chord should be part of the chord* (Eq. 4).

CounterStep
$$= \frac{\sum_{i=0}^{n-1} [\text{IsLeap}(i, i+1)(\text{PreCounterStep}(i, i+1) + \text{PostCounterStep}(i, i+1))]}{leapsN} \tag{2}$$

ChordOnLeap
$$= \frac{\sum_{i=0}^{n-1} [\text{IsLeap}(i, i+1)(\text{BelongsToChord}(i) + \text{BelongsToChord}(i+1))]}{leapsN} \tag{3}$$

$$\text{FirstNoteOnChord} = \frac{\sum_{i=0}^{n} (\text{IsFirstNote}(i) \times \text{BelongsToChord}(i))}{chordsN} \tag{4}$$

First, we remind the reader that the definition of an interval in music theory is the **difference between two pitches**. In Western music, intervals are mostly differences between notes belonging to the diatonic scale (for example, considering a $Cmajor$ key, the interval between C and D is a *second*, the interval between C and E is a *third* and so on).

To clarify what *counter step-wise motion* means: if we examine a leap of a fifth from C to G as in Fig. 3 (assuming we are in a $Cmajor$ key), this is a upward movement from a lower note to a higher one, a counter step-wise approach would mean that the C would be preceded by a higher note (creating a downward movement) with an interval of a second, so a D. Likewise following

Fig. 3. Example of counter step-wise approach and departure from a leap (C-G).

the leap in a counter step-wise motion would mean that we need to create a downward movement of a second after the G, so we need an F to follow.

The reason we introduce this objective is that this simple technique makes leaps much easier on the listener's ear, otherwise they often sound too abrupt by suddenly changing the range of notes the melody is playing. The PreCounterStep and PostCounterStep functions are boolean functions that respectively check if the note preceding and following the leap approaches or departs with a contrary interval of a second.

The reason for having the leap notes – the two notes that form a leap bigger than a second – be part of the underlying chord is that leaps are intrinsically more interesting than a step-wise motion, this means that the listener unconsciously considers them more meaningful and pays more attention to them. If these leaps contain notes that have nothing to do with the underlying chord, even if they do not present real dissonances, they will be perceived as dissonant because they create unexpected intervals with the chord notes. Trying to include them as part of the chord gives a better sense of coherence that the listener will consider as pleasant. The last objective simply underlines the importance of the first note after a chord change, by playing a note that is part of the chord we reinforce the change and make it sound less discordant.

Note that these objectives, by the nature of multi-objective optimization, will generally not all be completely satisfied. This is fine, because satisfying all objectives might make the generated music sound too mechanical, while these are "soft" rules that we want to enforce only to a certain point (contrary to the constraints of the infeasible population, which always need to be satisfied).

5.3   Accompaniment Generation

Accompaniment is included in the composition because, not only chords and melody give identity to music, but also rhythm. The accompaniment is divided into two parts: a **basic rhythm** (a collection of note duration) and a **basic note progression** (an *arpeggio*). We can progress from the accompaniment representation to a score of the accompaniment by creating notes with duration from the basic rhythm and pitches from the progressions (offset on the current underlying chord).

Accompaniments are generated through a stochastic process involving combinations and modifications (inversions, mutations, etc.) of some elements taken from a small archive of basic rhythms. Specifically we have four basic rhythm patterns and two basic arpeggios (see Figs. 4 and 5).

Fig. 4. Basic rhythms.

Fig. 5. Basic arpeggios. These are represented as if they were played under a C major or C minor chord, but are transposed depending on what chord they appear underneath. Also the rhythmic notation of the arpeggio is dependent on the rhythmic structure.

The algorithm performs the following steps:

1. choose a basic rhythm and basic arpeggio;
2. shuffle the elements of the basic rhythm;
3. shuffle the elements of the arpeggio;
4. increase the basic rhythm: with a probability of 0.5, either the biggest (longest duration) or a random element is split in two elements of half the size of the original duration. This function is called recursively with linearly decreasing probability;
5. increase the arpeggio to match the new size of the basic rhythm: this is done by introducing at a random index of the arpeggio a new random pitch that already belongs to the arpeggio.

The rhythm presented in the final music will be modified by the real-time affective music composer for variety or for affect expression, while still maintaining a rhythmic and harmonic identity that will be characteristic of the composition.

6   Experiment Design

We performed an extensive quantitative study to validate our music generation approach. The main objective is to investigate the contribution of each component of the framework to the quality of the music created. To do this, we systematically switched off components of our generator and replaced them with random generation. From these random "broken" compositions and the complete algorithm we created various pair-wise samples to test against each other. (This method was inspired by the "ablation studies" performed by e.g. Stanley [24].) As the quality of music is a subjective matter, we conducted a survey where participants are asked to prefer one of two pieces of music presented to them (one generated by the complete algorithm and one from a "broken" generator with one component disabled) and evaluate them according to four criteria: *pleasantness, randomness, harmoniousness* and *interestingness*. Using these four criteria presents a good overview of the preference expressed by the participant. Note that no definition of these terms is offered in the survey, so we have no guarantee they might not be interpreted differently by individual test subjects.

Pleasantness measures how pleasing to the ear the piece is, but this alone is not sufficient to describe the quality of the music produced. There are countless pieces of music that do not sound pleasant, but may nonetheless be considered by the listener as "good" music. In fact, in music, it is of course common to introduce uncommon (and even discordant) chord sequences or intervals to express different things like affect, narrative information and other effects. Also note that some alterations or passages can be specific of a music style. Moreover, discordant intervals is more and more accepted to the ear the more repeated they are (see dodecaphonic music, for example).

Interestingness is introduced to overcome the just described limitations of the pleasantness criteria: in this way we are able to test if one of our "broken" versions might introduce something that would result in something considered interesting to the listener, even when the composition is not as pleasant or harmonic. Note that this is a very subjective measure, as most people would find different things interesting.

On the other hand, *harmoniousness* might be confused with pleasantness, but we hope that it will be seen as a more objective measure: less of a personal preference and more of an ability to recognize the presence of dissonances and harmonic passages.

Finally, *randomness* gathers a measure of how structured the music appears to the listener. It is not only a measure of dissonance (or voices being off-key), but also of how much the music seems to have a cohesive quality and internal structure. Examples of internal structure are: (i) voices working together well (ii) coherent rhythmic structure (iii) the chord sequence presents tension building and resolution.

An online survey was developed with HTML and PHP, using a MySQL database to hold the data collected. Participants were presented with pairs-wise music clips and asked to evaluate them using the previously described four criteria. Each criteria has a multiple choices question structured as:

Which piece do you find more pleasing? *"Clip A"/"Clip B"/"Neither"/ "Both Equally"*

Where the last word (e.g. "pleasing") is dependent on the criteria. We also include the more neutral answers "Neither" and "Both Equally" to avoid randomness in the data from participants who cannot decide which clip satisfies the evaluation criteria better or worse. Other benefits of doing this are: avoiding the participant getting frustrated and giving us some possibly interesting information on cases where the pieces are considered equally good/bad.

Note that for the first five questions in the survey instrument the pair-wise clips always included one clip from the complete generator. After five trials, the clip pairs are picked at random between all the groups. In this way, we hoped to collect enough data to be able to make some observations between the four "broken" generators. The motivation behind this design choice is that our main question is evaluating the complete generator against all possible alternatives, so attention to the complete architectural music generator has priority. This also has a practical justification in the fact that, with the number of groups we have (five), testing all possible combinations and gather enough data would be near impossible. The survey has no pre-defined end: the user is able to continue answering until he/she wants, and can close the online survey at any time or navigate away from it without data loss. In the preamble, we encouraged participants to do at least five comparisons.

6.1   Music Clip Generation

The five groups were examined (as dictated by the architecture of our generator):

A. Complete generator: normal composition generator as described in Sect. 5;
B. Random chord sequence: the chord sequence module is removed and replaced with a random selection of chords;
C. Random unconstrained melody: the melody generation evolutionary algorithm is replaced with a random selection between all possible notes in the melody range (two octaves);
D. Random constrained melody: the melody generation evolutionary algorithm is replaced with a random selection between all possible notes belonging to the key of the piece in the melody range (two octaves). We decided this was necessary as (by design) our melody evolution approach is restricted to a diatonic context;
E. Random accompaniment: the accompaniment generation is replaced by a random accompaniment abstraction (we remind the reader that an accompaniment abstraction is defined by a basic rhythm and note sequence).

For each of these 5 groups, 10 pieces of music are created. For the sake of this experiment the affect expression has been kept to a neutral state for all the groups and we used the same algorithms to improvise on the composition abstraction. There is therefore no test of the music generators affect generation but rather of the music quality form the complete architecture compared to the architectural alternatives. The clips for the various groups can be accessed at http://msci.itu.dk/evaluationClips/

7   Results and Analysis

The data collected amounts to a total of 1,291 answers for each of the four evaluation criteria from 298 participants. Of the survey trials generated, 1,248 contained a clip generated with our complete music generator algorithm (A). Table 1 shows how many responses were obtained for each criteria and how many neutral answers were collected.

For now we only consider definitive answers (the participant chooses one of the music clips presented), we will look at the impact of the neutral answers at the end of this section. Under this constraint, the data becomes boolean: the answers are either *"user chooses the clip from the complete generator (A)"* or *"user chose the clip from the broken generator (B-E)"*. To analyze this data we use a two-tailed binomial test, which is an exact test of the statistical significance of deviations from a theoretically expected random distribution of observations into two categories. The null hypothesis is that both categories are equally likely to occur and, as we have only two possible outcomes, that probability is 0.5.

7.1 Complete Generator Against All Other Groups

Firstly, let us consider the combined results of all the "broken" groups (B-D) against the complete generator (A): as can be seen from Table 1, we have statistically highly significant differences for the *pleasing*, *random* and *harmonious* categories, while we have a p-value of 0.078 for the *interesting* category. This means that we can refuse the null hypothesis and infer a difference in distribution between choosing the music generated by the complete algorithm (A) and the "broken" ones (B-E).

We can affirm that our generator (A) ranked better than all the other versions (B-E) for three of our four criteria, with the exception of *interestingness*, where there is no statistically significant difference. Interestingness is clearly a very subjective measure, and this may explain the result. Moreover, examining the ratio of neutral answers obtained for this criteria, it can be inferred that it is almost 26 %, so a much higher neutral response that for the other criteria. This suggests that in a higher number of cases participants could not say which composition they found more interesting. A possible explanation is that, as the affect expression (which also includes musical features such as tempo and intensity) is held in a neutral state, equal for all pieces, after hearing a number of clips the listener does not find much to surprise him/her. Also the duration of the generated pieces (30 s) might not allow sufficient time to determine interestingness.

Table 1. Amounts of correct, incorrect and neutral answers to our criteria for the complete generator against all the "broken" generators (B-E) combined. Keep in mind that in the case of the random criteria the listener is asked to select the clip that he/she feels the most random, so it is entirely expected that a low number of participants choose the random clip (E) against the complete generator (A).

Choice	Pleasing	Random	Harmonious	Interesting
Choose the complete generator (A)	654	197	671	482
Choose a "broken" generator (B-E)	240	633	199	327
A neutral answer	197	261	221	282
Total non-neutral answers	894	830	870	809
Binomial test p-value	7.44E-21	2.75E-77	2.05E-29	7.81E-02

Table 2. Answers and results of the binomial test for pairs comprised of the full generator and the one with random chord sequences.

A versus B	Pleasing	Random	Harmonious	Interesting
Successes	121	71	112	98
Failures	93	117	84	98
Totals	214	188	196	196
Binomial test p-value	3.23E-02	4.90E-04	2.68E-02	5.28E-01

Table 3. Answers and results of the binomial test for pairs comprised of the full generator (A) and the one with unconstrained random melody (C).

A versus C	Pleasing	Random	Harmonious	Interesting
Successes	221	21	236	144
Failures	26	221	19	72
Totals	247	242	255	216
Binomial test p-value	5.15E-40	1.44E-43	4.11E-49	5.46E-07

7.2 Complete Generator Against Random Chord Sequence Generation

If we only consider the pairs that included the complete generator (A) and the one with random chord sequences (B) (Table 2) we, again, obtain statistically significant differences in the distribution of the answers for the *pleasing, random* and *harmonious* criteria. In this case we have a very high p-value for *interestingness* (more than 0.5), in fact we have the same amount of preference for the complete generator (A) and the "broken" one (B). We can explain this by considering that the disruptive element introduced by this modification of the algorithm is mitigated by the fact that the rest of the system tries to create as pleasing music as it can based on the chord sequence given. So, for most of the time, the music will not have notes that sound out of key or that do not fit well with the chord sequence. Still, we can observe how the listener is capable of identifying that, while the piece does not sound discordant or dissonant, it lacks the structure of tension-building and tension-releasing. This explains how the complete generator (A) is preferred for all other criteria. It is interesting to note how the act itself of presenting the listener with uncommon chord sequences does create an increase in the interestingness of the music.

7.3 Complete Generator Against Unconstrained Melody Generation

When we consider the unconstrained melody group we have statistically significant differences for all criteria, with some extremely strong significance (Table 3). These results are as we expected, as the melody plays random notes that conflict with both the chord sequence and the accompaniment.

7.4 Complete Generator Against Constrained Melody Generation

The results given by the constrained random melody generation (D) are more interesting (Table 4). First, we notice no statistically significant values for the *pleasing* and *interesting* criteria. This is explained by the fact that the melody never goes off key, so it never presents off-key notes and never sounds abruptly "wrong" to the listener's ear. Yet, the *random* and *harmonious* criteria are statistically significant. Remembering how we described these criteria we notice that the more objective criteria (*random* and *harmonious*) are those that demonstrate a difference in distribution. We believe this reinforces how, although compositions made in this group never achieve a bad result, the listener is still able to identify the lack of structure (randomness) and lack of consideration of the underlying chords of the melody (harmoniousness). An example of the first case would be a melody that jumps a lot between very different registers; this would make the melody sound more random than the melodies we evolve using (A), which follow more closely the guidelines of a singing voice. Harmoniousness can be influenced by the fact that, over a chord (expressed by the accompaniment), the melody can play notes that create intervals that 'muddle' the clarity of the chord to the listener's ear.

7.5 Complete Generator Against Random Accompaniment Generation

Finally, for the last group, the random accompaniment generation (E), gives us very clear statistically significant results on all criteria (Table 5). A lot of the harmony expression depends on the accompaniment generation, and when this is randomized it is no wonder that the piece sounds confusing and discordant. This is reflected in the trial data.

Table 4. Answers and results of the binomial test for pairs comprised of the full generator (A) and the one with constrained random melody (D).

A versus D	Pleasing	Random	Harmonious	Interesting
Successes	125	81	120	108
Failures	100	109	85	94
Totals	225	190	205	202
Binomial test p-value	5.47E-02	2.49E-02	8.68E-03	1.80E-01

Table 5. Answers and results of the binomial test for pairs comprised of the full generator (A) and the one with random accompaniment (E).

A versus E	Pleasing	Random	Harmonious	Interesting
Successes	188	25	203	132
Failures	21	186	12	63
Totals	209	211	215	195
Binomial test p-value	5.00E-35	6.58E-32	3.01E-46	4.35E-07

8   Conclusion and Future Work

This paper describes a new component-based system for music generation based on creating an abstraction for musical structure that supports real-time improvisation. We focus on the method of creating the abstractions ("compositions"), that consists of the sequential generation of (i) chord sequences, (ii) melody and (iii) an accompaniment abstraction. Our novel approach is described to evolve melody in detail: non-dominated sorting with two feasible-infeasible populations genetic algorithm (NSFI-2POP).

Returning to the main question of the paper, (*do all parts of the music generation system add to the music produced?*), we have described an evaluation in which we created music with our generator substituting various components with randomized generators. In particular we observed four broken groups: *random chord sequences, random melody constrained* (to the key of the piece), *random melody unconstrained* and *random accompaniment*. An evaluation of the music clips generated by these "broken" versions was compared to music clips created by the complete algorithm according to four criteria: *pleasantness, randomness, harmoniousness* and *interestingness*.

Analysis of the data supports the assertion that participants prefer the complete system in three of the four criteria: (*pleasantness, randomness* and *harmoniousness*) to the alternatives offered. The results for the *interestingness* criteria are however not definitive, but suggest that some parts of our generator have a higher impact in this criteria. It is also noteworthy that there is no statistical significant difference between preferences between *constrained melody group* (C) and the *complete generator* (A) for the *pleasantness* criteria.

Future work will focus on developing and evaluating the affect-expression capabilities of the system, we will probably follow the methodology described by Scirea *et al.* for characterizing control parameters through crowd-sourcing [25]. In summary, we show (i) how each part of our music generation method assists creating music that the listener finds more pleasant and structured and (ii) presented a novel GA method for constrained multi-objective optimization.

References

1. Papadopoulos, G., Wiggins, G.: Ai methods for algorithmic composition: a survey, a critical view and future prospects. In: AISB Symposium on Musical Creativity, Edinburgh, UK, pp. 110–117 (1999)
2. Konečni, V.J.: Does music induce emotion? A theoretical and methodological analysis. Psychol. Aesthetics Creativity Arts **2**(2), 115 (2008)
3. Yannakakis, G.N., Togelius, J.: Experience-driven procedural content generation. IEEE Trans. Affect. Comput. **2**(3), 147–161 (2011)
4. Miranda, E.R.: Readings in Music and Artificial Intelligence, vol. 20. Routledge, New York (2013)
5. Wooller, R., Brown, A.R., Miranda, E., Diederich, J., Berry, R.: A framework for comparison of process in algorithmic music systems. In: Generative Arts Practice 2005 – A Creativity & Cognition Symposium (2005)

6. Robertson, J., de Quincey, A., Stapleford, T., Wiggins, G.: Real-time music generation for a virtual environment. In: Proceedings of ECAI-98 Workshop on AI/Alife and Entertainment, Citeseer (1998)
7. Smaill, A., Wiggins, G., Harris, M.: Hierarchical music representation for composition and analysis. Comput. Humanit. **27**(1), 7–17 (1993)
8. Loughran, R., McDermott, J., O'Neill, M.: Tonality driven piano compositions with grammatical evolution. In: IEEE Congress on Evolutionary Computation (CEC), pp. 2168–2175. IEEE (2015)
9. Dahlstedt, P.: Autonomous evolution of complete piano pieces and performances. In: Proceedings of Music AL Workshop, Citeseer (2007)
10. Miranda, E.R., Biles, A.: Evolutionary Computer Music. Springer, London (2007)
11. Rigopulos, A.P., Egozy, E.B.: Real-time music creation system, US Patent 5,627,335, 6 May 1997
12. Meier, S.K., Briggs, J.L.: System for real-time music composition and synthesis, US Patent 5,496,962, 5 March 1996
13. Livingstone, S.R., Brown, A.R.: Dynamic response: real-time adaptation for music emotion. In: Proceedings of the 2nd Australasian Conference on Interactive Entertainment, pp. 105–111 (2005)
14. Brown, D.: Mezzo: an adaptive, real-time composition program for game soundtracks. In: Proceedings of the AIIDE 2012 Workshop on Musical Metacreation, pp. 68–72 (2012)
15. Zitzler, E., Deb, K., Thiele, L.: Comparison of multiobjective evolutionary algorithms: empirical results. Evol. Comput. **8**(2), 173–195 (2000)
16. Deb, K., Pratap, A., Agarwal, S., Meyarivan, T.: A fast and elitist multiobjective genetic algorithm: NSGA-II. IEEE Trans. Evol. Comput. **6**(2), 182–197 (2002)
17. Kimbrough, S.O., Koehler, G.J., Lu, M., Wood, D.H.: On a feasible-infeasible two-population (FI-2Pop) genetic algorithm for constrained optimization: distance tracing and no free lunch. Eur. J. Oper. Res. **190**(2), 310–327 (2008)
18. Deb, K., Pratap, A., Meyarivan, T.: Constrained test problems for multi-objective evolutionary optimization. In: Zitzler, E., Deb, K., Thiele, L., Coello Coello, C.A., Corne, D.W. (eds.) EMO 2001. LNCS, vol. 1993, pp. 284–298. Springer, Heidelberg (2001)
19. Chafekar, D., Xuan, J., Rasheed, K.: Constrained multi-objective optimization using steady state genetic algorithms. In: Cantú-Paz, E., et al. (eds.) GECCO 2003. LNCS, vol. 2723, pp. 813–824. Springer, Heidelberg (2003)
20. Jimenez, F., Gómez-Skarmeta, A.F., Sánchez, G., Deb, K.: An evolutionary algorithm for constrained multi-objective optimization. In: Proceedings of the Congress on Evolutionary Computation, pp. 1133–1138. IEEE (2002)
21. Isaacs, A., Ray, T., Smith, W.: Blessings of maintaining infeasible solutions for constrained multi-objective optimization problems. In: IEEE Congress on Evolutionary Computation, pp. 2780–2787. IEEE (2008)
22. Mugglin, S.: Chord charts and maps. http://mugglinworks.com/chordmaps/chartmaps.htm. Accessed 14 October 2015
23. Deb, K.: Multi-objective Optimization Using Evolutionary Algorithms, vol. 16. Wiley, Chichester (2001)
24. Stanley, K.O., Miikkulainen, R.: Evolving neural networks through augmenting topologies. Evol. comput. **10**(2), 99–127 (2002)
25. Scirea, M., Nelson, M.J., Togelius, J.: Moody music generator: characterising control parameters using crowdsourcing. In: Johnson, C., Carballal, A., Correia, J. (eds.) EvoMUSART 2015. LNCS, vol. 9027, pp. 200–211. Springer, Heidelberg (2015)

'Turingalila' Visual Music on the Theme of Morphogenesis

Terry Trickett[✉]

London, UK
terrytrick@btinternet.com

Abstract. Alan Turing's paper 'The Chemical basis of Morphogenesis', written in 1952, is a masterpiece of mathematical modelling which defines how self-regulated pattern formation occurs in the developing animal embryo. Its most revolutionary feature is the concept of 'morphogens' that are responsible for producing an almost limitless array of animal and fish markings. Turingalila, a piece of Visual Music, takes morphogenesis as its theme. The diversity of forms evident in my projected images are based on just two Turing Patterns which are 'perturbed' to reveal processes of self-organisation reminiscent of those found in nature. A live performance of Turingalila forms the focal point of my oral presentation. It is prefaced by an examination of how artistic potential has been unleashed by Turing's biological insights and concludes with comments on how Turing's ideas are exerting an ever increasing impact in today's world.

Keywords: Alan Turing · Morphogenesis · Morphogens · Turing Patterns · Visual Music · Germaine Tailleferre · Clarinet · Biological mathematics

1 Introduction

'Turingalila' is a piece of biologically inspired Visual Music – my most recent adventure in this art form which, for me, combines projected visual images with live musical performance. In this paper I explain the derivation of my visual images and how they relate to my chosen music. But first, as an introduction, I am tracing the source of my biologically inspired idea and giving some account of how I see artistic potential being unleashed by revolutionary biological thinking.

It was only towards the end of his life that Alan Turing turned his attention to the subject of Morphogenesis. Ever since his school days he had shown an interest in the shape, form and growth of plants and flowers but, over time, this interest extended from botanical growth and form to embrace biological species. He attacked the problems posed by the life sciences with extreme care and with considerable humility and brought to their study an intuition that was more mathematical than biological. The result, his theory of morphogenesis, provided a paradigm shift in our way of thinking that, ever since, has continued to stimulate considerable controversy and experimental work.

Link to video of Turingalila https://youtu.be/H9jAdbfJCB0.

C. Johnson et al. (Eds.): EvoMUSART 2016, LNCS 9596, pp. 218–224, 2016.
DOI: 10.1007/978-3-319-31008-4_15

As in Turing's innovative work on computer design, his researches into mathematics and his consideration of 'Thinking Machines', he brought the mind of a genius to bear on the subject of biology. Computation played a part in his endeavour because much of the arithmetic work was carried out on the Ferranti Mark 1 computer, located in Manchester. Turing had written the guide for programmers and was able personally to prepare the perforated magnetic tape required to feed instructions into the machine. Given these comparatively primitive tools and techniques, Turing was able to produce a morphogenetic thesis of extreme sophistication. It has opened up a model on the life sciences that has proved to be extraordinarily successful; further, it has provided the source of inspiration for Turingalila! The result of applying the model, as I will show, consistently produces images which mimic those found in nature and it is with these that I weave a constantly changing pattern of growth and form.

2   The Magic of Morphogenesis

Turing's motivation in tackling morphogenesis was generated by his conviction that complex biological growth and development could be achieved via simple natural processes. To quote his own paper on the subject, his purpose was "to discuss a possible mechanism by which the genes of a zygote may determine the anatomical structure of the resulting organism [1]". From the beginning he was certain that well-known physical laws would be sufficient to account for many of the facts. The title of his paper *The Chemical Basis of Morphogenesis* reveals his line of thought which, at the time, 1952, was both audacious and innovative. The systems he considered consisted of "masses of tissue which are not growing, but within which certain substances are reacting chemically, and through which they are diffusing". He called these substances morphogens, with the intention that the word would convey the idea of form producer. Turing Patterns result from the reaction - diffusion system that he proposed. They occur in conditions where "the right rates for the chemical reaction are matched by the right diffusion rates of reacting species [2]".

The first experimental observations of Turing Patterns occurred nearly 40 years after Turing had submitted his paper to the Royal Society. During this long period of inaction, his ideas were largely ignored or, at best, greeted with considerable scepticism. Limited investigations into chemical reaction - diffusing systems focused on skin patterns in mammals and fish; scientists discovered that Turing Patterns, showing arrangements of stripes and spots, arose spontaneously from random initial conditions (Fig. 1).

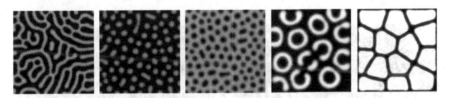

Fig. 1. Two dimensional patterns generated by the Turing model. These patterns were made by an identical equation with slightly difference parameter values. Images courtesy of Kondo and Miura [3].

Turing, himself, had posed two key questions: "How can a single fertilised cell grow and develop into a fully fledged organism? How can complex biological growth and development be achieved via simple, natural mechanisms? [1]" The complete answers to these questions are still not known but there is no doubt that Turing's initial hypothesis has succeeded in moving the science of morphogenesis many steps forward. During on-going investigations, it is the simulation of patterns in animal, fish and insect markings that has continued to arouse special interest and comment from both scientists and lay audiences (Fig. 2).

Fig. 2. These patterns are typical of the animal, fish and insect markings that have been tested against Turing's reaction - diffusion model.

To summarise progress since Turing over the last 60 years or so, the reaction - diffusion (RD) model has succeeded in revealing how self-regulated pattern formation occurs in the developing animal embryo. As a result, Turing's proposal is now recognised as a masterpiece of mathematical modelling which explains how spatial patterns develop autonomously. Its most revolutionary feature is the concept of a 'reaction' that produces morphogens. These 'form producers' can generate a nearly limitless variety of spatial patterns. "The intricate involutions of seashells, the exquisite patterns of feathers, or the breathtaking diverse variety of skin patterns can all be modelled within the framework of the Turing model [3]". As Shigeru Kondo and Takashi Miura conclude in their paper, *Reaction - Diffusion Model as a Framework for Understanding Biological Pattern Formation,* the remarkable similarity that has been observed between simulated experiments and reality strongly points to Turing mechanisms as an underlying principle of biological patterning (Fig. 3).

However, scepticism towards Turing's hypothesis remains in the minds of some biologists and embryologists; they claim that no matter how vividly or faithfully a mathematical simulation might replace an actual biological pattern, this does not consti-tute proof that the simulated state reflects reality. To put these views into perspective, it can be argued that Turing's amazing prescience in devising his RD model has deter-mined a successful method for uncovering a myriad of Turing Patterns that match the

Fig. 3. Simulation images of fish patterns – courtesy of Sanderson *et al.* [4].

natural world but, whether or not this system can be extended to the more complex forms of bodies and anatomies remains a matter of conjecture and much ongoing research – see Conclusion: Patterns of Change below.

3   Creating 'Turingalila' as Visual Music

First, I should explain that my title is a pun on the word 'Turangalîla' from Olivier Messiaen's *Turangalîla Symphonie*. The composer derived his title from two Sanskrit words, *turanga* and *lîla*, which, roughly translated, mean 'love, song, and hymn of joy, time, movement, rhythm, life and death' – in fact everything embraced by the concept of morphogenesis!

In creating the work, I have taken full advantage of advances in computer generated imagery which now provide opportunities to imitate the irrepressible impulse of nature to create patterns. "Nature is not the austere censor, as biologists used to believe, stripping away all unnecessary detail and ornamentation [5]". Some patterns just happen without regard to any specific function. It is at this level that computer simulations of the results of the RD system can elucidate, delight, surprise and entertain the viewer. This is my aim in creating 'Turingalila'; I cannot simulate the dynamics of the RD system itself but I can demonstrate the visual ramifications of Turing's complex mathematics in a manner which might add to people's general understanding of his achievement.

My approach has been to take just two Turing Patterns – a hexagonal stripe pattern and a hexagonal spot pattern (Fig. 4) and then to perturb these patterns in all sorts of ways to observe a process of continuous change and regeneration (Figs. 5 and 6). I use a varied set of Processing codes to achieve this end and, as ever, choose image sequences that appear, to me, to provide convincing illustrations of growth and form. This imagery then needs to be matched against music which, similarly, develops continuously from start to finish. For this purpose, I have selected Germaine Tailleferre's Sonata for Solo Clarinet. It's one of the few compositions where Taillefere experimented with serialism although, in the end, she gave only a nod in the direction of twelve tone writing. Her heart wasn't in it; when you listen to her Sonata you hear only her inimitable style which remained predominately neoclassical and lyrical.

Fig. 4. Two Turing Patterns – a hexagonal stripe pattern and a hexagonal spot pattern

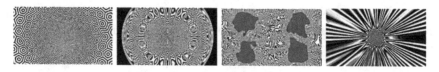

Fig. 5. 'Perturbances' applied to Turing's hexagonal stripe pattern

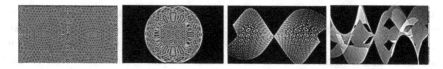

Fig. 6. 'Perturbances' applied to Turing's hexagonal spot pattern

The patterns of Tailleferre's notes, in her Sonata, reflect, to my mind, the Turing Patterns which form the basis of my visual imagery. In the first movement, *allegro tranquillo,* the imagery is based on a simple presentation of the hexagonal stripe pattern (Fig. 7). This evolves and changes shape as a result of 'perturbances' that I introduce to produce, at its conclusion, a series of sinuous arabesques. The second movement, *andantino expressivo,* weaves a series of perturbed images based on the hexagonal spot Turing Pattern (Fig. 8). Somewhat surprisingly, animal and fish forms can be seen emerging from the resulting constantly moving shapes. In other words, the Turing Patterns do appear to produce elements of complex biological growth and development which extend beyond the mere patterning of pigmented surfaces. The last movement, *Allegro brioso,* overlays both of my selected patterns, one on top of the other (Fig. 9). Here, too the results provide a few hints of mammalian movement.

Fig. 7. The first movement of Turingalila is based on a simple presentation of Turing's hexagonal stripe pattern. At its conclusion it reveals a series of sinuous arabesques.

Fig. 8. The second movement is based on Turing's hexagonal spot pattern. It reveals animal and fish forms emerging from a series of constantly moving shapes.

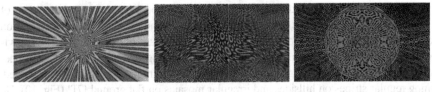

Fig. 9. In the third movement, the Turing hexagonal stripe and hexagonal spot patterns are combined, one on top of the other. Some hints of mammalian movement are revealed.

4 Conclusion: Patterns of Change

In creating Turingalila, I've found meaning in my shapes and edited my results to produce Visual Music. During the process, as you have seen, an extraordinary diversity of forms has emerged seemingly spontaneously from image sequences generated by Processing codes. It's a process of self-organisation, generated with a set of remarkably simple rules, which produces shapes that are reminiscent of those found in nature. It seems that by interpreting Turing's ideas as a form of artistic expression I've, also, as it happens, hit upon a purposeful tool for understanding natural processes. All I'm doing, of course, is following in the footsteps of Alan Turing; my project relies on his mathematical and scientific brilliance. It seems to me that he has succeeded, posthumously, in his ambition to discover how complex biological growth and development can be achieved via simple natural mechanisms. But not everyone agrees; scepticism towards Turing's hypothesis remains in the minds of some scientists. Even as a non-scientist, I can see their point of view; it appears to defy common sense that mathematics can solve fundamental problems in biology. For this reason, it is only through hard-nosed scientific proof that doubts and scepticism towards Turing's theory can be dispelled. And, fortunately, this now is beginning to emerge:

Dr. Jeremy Green is a reader in developmental cell biology at King's College, London. His research lab analyses how cells build up tissues, specifically examining the molecules that give cells specific directions in three-dimensional space. His team discovered that the ridges on the roof of the mouth act like the stripes in Turing's theory. They identified both the morphogens and the anti-morphogens involved in the process and, by perturbing them in mice, they produced exactly the results that Turing's theory predicts. This is significant because the success of this experiment, and others like it, give promise that one day it will become possible to use morphogens to repair complex structures, rather than just patch small holes with a few cells. Jeremy Green gives full credit to Alan Turing's theory for providing the key to his own discoveries: "….it's the scientific explanation for the spontaneous miracle of development before birth …..the way morphogens drive hundreds of simple steps that make one part of an embryo different from another [6]". Other researchers are focusing on how Turing's insights can explain the way tubes within an embryo's developing chest split over and over again to create delicate bronchial lungs. "Even the regular array of teeth in our jaws probably got there by Turingesque patterning [6]".

If further proof were needed, Turing Patterns can now be found in many other areas of life – in weather systems, in the constellations of galaxies and the distribution of vegetation across the landscape. What's more, they can provide the means of extracting crucial knowledge from environmental data. Take vegetation patterning as an example. Christopher Klausmeier has noted that "vegetation in many semi-arid regions is strikingly patterned, forming regular stripes on hillsides and irregular mosaics on flat ground [7]" (Fig. 10). He concludes that these regular patterns result from a Turing-like instability and, further, that "the system provides a clear example of how non-linear mechanics can be important in detecting the spatial structure of plant communities." Another researcher, Jonathan Sherratt, has observed that "ecological pattern formation at the level of whole ecosystems is a new, exciting and rapidly growing research area, whose study is influenced strongly by Turing's ideas [8]". The importance of this on-going work will be immediately evident. In nature, changes in vegetation patterning take place over years or decades but, through the application of the Turing RD model, it becomes possible to foresee the future consequences of change ahead of time. Armed with this knowledge, steps can be taken to reverse any causes of detrimental change so as to ensure, for instance, the on-going and long term fertility of the landscape. In a future that is becoming increasingly uncertain, it will become ever more vital that patterns of change can be predicted well in advance. Turing Patterns enable us to do just that!

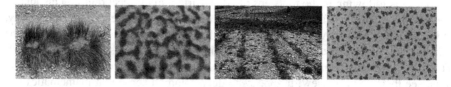

Fig. 10. Vegetation patterning in various semi-arid regions reveal tell-tale signs of Turing Patterns.

References

1. Turing, A.M.: The chemical basis of morphogenesis. Philos. Trans., R. Soc. B **237**, 37–72 (1952)
2. Quotation from Irving Epstein, Department of Chemistry and Center for Complex Systems, Brandeis University. http://www.wired.com/2011/02/turing-patterns/ (retrieved)
3. Kondo, S., Muira, T.: Reaction-diffusion model as a framework for understanding biological pattern formation. Science **329**, 267–275 (2010)
4. Sanderson, A.R., Kirby, R.M., Johnson, C.R., Yang, L.: Advanced reaction-diffusion models for texture synthesis. J. Graphics GPU Game Tools **11**, 47 (2006)
5. Ball, P.: Shapes, (Written on the Body), p. 182. Oxford University Press, Oxford (2009)
6. Arney, K.: How the zebra got its stripes, with Alan Turing. Mosaic, The Wellcome Trust, August 2014
7. Klausmeier, C.A.: Regular and irregular patterns in semiarid vegetation. Science **289**, 1826–1828 (1999)
8. Sherratt, J.A.: Turing Patterns in Deserts. Heriot-Watt University, Edinburgh. http://www.macs.hw.ac.uk/~jas/paperpdfs/sherratt-cie2012.pdf (unpublished paper)

Fitness and Novelty in Evolutionary Art

Adriano Vinhas[1]([✉]), Filipe Assunção[1], João Correia[1], Aniko Ekárt[2],
and Penousal Machado[1]

[1] Department of Informatics Engineering, CISUC, University of Coimbra,
Coimbra, Portugal
{avinhas,fga,jncor,machado}@dei.uc.pt
[2] Aston Lab for Intelligent Collectives Engineering (ALICE), Computer Science,
Aston University, Birmingham, UK
a.ekart@aston.ac.uk

Abstract. In this paper the effects of introducing novelty search in evolutionary art are explored. Our algorithm combines fitness and novelty metrics to frame image evolution as a multi-objective optimisation problem, promoting the creation of images that are both suitable and diverse. The method is illustrated by using two evolutionary art engines for the evolution of figurative objects and context free design grammars. The results demonstrate the ability of the algorithm to obtain a larger set of fit images compared to traditional fitness-based evolution, regardless of the engine used.

Keywords: Novelty search · Evolutionary art · Multi-objective optimisation

1 Introduction

Computational Creativity research posits that an answer to a problem is considered creative if it is both useful and novel[1] [1]. Although this definition is not consensual [2], it is the one that better fits the scope of this work. Bearing this in mind, and applying it to images, it was proved that expression-based evolutionary art, introduced by Sims [3], is theoretically able to generate any possible image [4,5]. In practice, however, the images generated by a given system tend to share the same overall appearance.

Previous works on evolving figurative or ambiguous images have focused on using an expression-based approach [6–9]. Although they achieved interesting results in terms of recognizability of the desired object(s), an important drawback revealed by the analysis of these works is that there are runs where the evolutionary algorithm is not able to evolve images that meet the requirement of resembling the desired object(s). Moreover, there are some runs which are able to generate those images, but the results tend to be very similar to each other, as a consequence of the evolutionary algorithm's convergence towards an

---

[1] In the context of the present paper, novelty means phenotypic diversity.

C. Johnson et al. (Eds.): EvoMUSART 2016, LNCS 9596, pp. 225–240, 2016.
DOI: 10.1007/978-3-319-31008-4_16

optimum. Problem deceptiveness or lack of explicit diversity exploration can be the explanation for these shortcomings.

When evolving Context Free Design Grammars (CFDGs)[10], the grammar representation expressiveness allows the creation of different individuals [11]. As such, CFDGs are capable of generating a family of solutions by combining a small set of production rules. However, during the evolutionary process, after a few generations individuals tend to converge towards an optimum, i.e., most individuals of the final population will share common genetic background and thus will tend to produce similar visual artifacts.

The work described here takes as its starting point these results and analysis, aiming to address the aforementioned shortcomings by proposing an algorithm that takes both novelty and fitness into account. Adopting the traditional definition of creativity, fitness stands for usefulness and phenotypic diversity stands for novelty. The algorithm proposed herein is used to evolve images in two different domains: figurative images and CFDGs. In both cases, when canonical fitness-based evolution is used, the last generations suffer from a lack of diversity. By incorporating novelty-based mechanisms into evolution we expect to increase diversity. The goal of this work is to propose a novelty search algorithm for evolutionary art purposes, and to analyse its capability to generate diverse and fit images, comparing, for each problem, the traditional fitness-based evolution with one that also takes novelty information into account.

The remainder of this paper is structured as follows. Section 2 surveys the state of the art regarding novelty search algorithms; Sect. 3 describes the proposed algorithm that will be applied to two different problems in Sects. 4 and 5. Experimental setup, results and their analyses are focused upon in Sect. 6. In Sect. 7 conclusions are drawn and future work is addressed.

2   State of the Art

Novelty search has become a trending topic in optimisation problems, where it has been used in fields such as robotics [12–14] and arts [15,16]. Even before the first formalisation of a novelty search algorithm, by Lehman and Stanley [17], there were already reports in the literature concerning the evolution of images using a novelty metric [18]. In the work of Saunders et al. [18], the novelty concept is mathematically modeled according to an interestingness definition based on two factors: (i) the ability to create artifacts out of the box (unexpectedness) and (ii) the feasibility of taking an action as a consequence of the discovery (actionability) [19]. This metric is then used in the context of an Evolutionary Algorithm (EA) designed to evolve novel imagery artworks. In essence, these abstract images were generated using an EA, in which the evaluation was performed by an agent in two steps. First, a self organizing map (SOM) [20] is used to determine the category where an image best fits. The novelty is computed as the classification error of an image being associated to the best category. The second step consists in applying the interestingness function, computing the interest of each image, given its novelty degree.

As mentioned before, the more recent concepts of novelty search follow the steps of Lehman and Stanley's work, which propose a novelty search algorithm that aims at valuing each image's uniqueness. For each individual, a novelty score is computed, taking into account its neighbours and an archive containing the most novel individuals. Each novelty score computation requires a phenotype comparison, using a dissimilarity metric, between the individual being evaluated and a set of neighbours chosen from the population and the archive. Then, the novelty score for the individual being evaluated (ind_{eval}) is defined as the average of the dissimilarity scores of the k most similar neighbours, as in:

$$\text{nov}(\text{ind}_{eval}) = \frac{1}{k} \sum_{j=1}^{k} \text{dissim}(\text{ind}_{eval}, \text{ind}_j), \tag{1}$$

where *dissim* denotes the chosen dissimilarity metric, and j the j-th most similar individual (ind_j) when compared to ind_{eval}.

In order to enable the use of novelty search techniques for constrained problems, Stanley and Lehman proposed the Minimal Criteria Novelty Search (MCNS) algorithm [21]. The idea of employing novelty search *per se* is not suitable for this kind of constrained problems because the function used to evolve individuals enhances diversity without any boundaries and hence, creates unsuitable solutions. Therefore, MCNS tries to tackle this problem by decimating the population, i.e., any valid solution would be evaluated using the respective novelty function and unsuitable solutions would be assigned zero fitness.

Liapis et al. based their work on MCNS [22]. Although MCNS does not favour the creation of unsuitable solutions, the idea of decimating a population means that one is assigning the same value (zero) to all unsuitable solutions. As the initial generations of EAs usually hold lots of unsuitable solutions, the selection process would become similar to a random search. This drawback becomes more evident if the problem that one is trying to solve is highly constrained. In order to tackle this disadvantage, Liapis et al. proposed two solutions based on the feasible-infeasible two-population genetic algorithm (FI-2pop GA) [23]. In essence, the idea of FI-2pop GA is to evolve valid solutions and unsuitable solutions as if they are two different populations. However, there can be exchanges of individuals between populations, because a valid solution can become unsuitable and vice-versa. The two solutions proposed were: (i) a feasible-infeasible novelty search (FINS), which consists in evolving the feasible population with novelty search as in the work of Lehman and Stanley [17], while evolving the infeasible one with the goal of minimising each individual's distance to the feasibility border, and (ii) a feasible-infeasible dual novelty search (FI2NS), which consists in applying novelty search as in [17], in both populations separately, using two different novelty archives. Their results suggest that both solutions are superior to the MCNS performance but, while FINS is able to get feasible individuals within a lower number of generations, FI2NS creates a more diverse set of feasible solutions.

Despite the good performance of novelty approaches, authors tend to point out their lack of ability to maintain good performance in problems with big search spaces. The rationale is that novelty approaches saturate the search space, until they eventually get a reasonable or optimal solution. However, in big search spaces this saturation is much harder to achieve, and the solutions will fail to reach an optimal solution or, in an optimistic scenario, the desired solution is found after a large number of generations. This problem motivated the development of hybrid mechanisms, which use both fitness and novelty-based functions to guide evolution. Within this scope, we highlight three works: [12,14,24].

In Krcah et al. [12], novelty search is applied during a predefined number of generations, and the remainder of the process is performed with a goal-based evolution. Although it is a simple solution, it will not avoid a later convergence to a local optima. In Cuccu et al. [24], novelty and fitness are merged using a weighted combination approach. This solution requires prior knowledge about the weight values, which depend on the domain. Finally, Mouret et al. [14] used a Pareto-based multi-objective evolutionary algorithm to combine fitness and novelty, in order to help a robot which is subject to a maze navigation task. Even though it is an interesting idea, preliminary tests indicate that it is not appropriate in an evolutionary art context since finding phenotypically diverse images is significantly easier than finding fit images and the search space is vast. Thus, in this conditions evolution will produce a seemingly endless stream of images that are "novel" but not fit.

3   Proposed Algorithm

In this section, a new novelty search algorithm is proposed, designed to evolve a diverse set of suitable images. Therefore, the main goal of this algorithm is to generate a more phenotypically diverse set of images than the set that would be created by a traditional fitness based EA. In essence, it is a method capable of evolving images according to two criteria that are chosen automatically by analysing the quality of the images produced in each generation. One criterion is to look for the best images according to a fitness function and the other consists in taking novelty and fitness as two different objectives to be maximised. As previously mentioned, considering fitness or novelty alone is not suitable for the problem at hand [25].

The algorithm's flowchart is similar to the traditional EA one, differing only in two main aspects: (i) the creation of an archive to store the most novel solutions and, (ii) a customised selection mechanism which is able to consider single or multiple objectives using a tournament based strategy. The algorithm's flow is shown in Fig. 1, and can be summarised as follows:

1. Randomly initialise the population;
2. Render the images (phenotypes) from the individuals' genotypes;
3. Apply the fitness function to the individuals;
4. Select the individuals that meet the criteria to be in the archive (archive assessment);

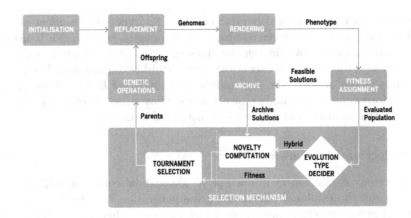

Fig. 1. Flow of the hybrid algorithm proposed.

5. Select the individuals to be used in the breeding process. The individuals are picked using one of the following criteria: (i) according to their fitness, as a standard EA; (ii) taking into account both the fitness and the novelty metric, computed using the archive members;
6. Employ genetic operators to create the new generation of solutions, that will replace the old one;
7. Repeat the process starting from step 2, until a stop criterion is met.

3.1 Archive Assessment

In this work, the archive has an unlimited size and it plays an important role, because it is used to evaluate our solution and prevents the algorithm from exploring areas of the search space already seen before. The idea is that the archive should represent the spectrum of images found to date, and for this reason, the bigger the archive is, the more capable is the algorithm of generating suitable and diverse images. Whereas in the previously mentioned works the archive size is limited, we opted for not restricting it.

At this stage, a candidate individual has its fitness assigned and it has to meet two requirements in order to be added to the archive: (i) its fitness must be greater or equal than a threshold f_{min}, defined by the user; (ii) it needs to be different from those that already belong to the archive. This process is performed by computing the average dissimilarity between the candidate and a set of k-nearest neighbours. When the average dissimilarity is above a predefined dissimilarity threshold, $dissim_{min}$, the individual is added to the archive. The dissimilarity metric for an image i is computed as:

$$\text{dissim}(i) = \frac{1}{\max_{arch}} \sum_{j=1}^{\max_{arch}} d(i,j), \qquad (2)$$

where max_{arch} is a predefined parameter which represents the number of most similar images to consider when comparing with image i, and $d(i, j)$ is a distance metric that measures how different two images (i and j) are. From this dissimilarity measure there are two exceptions that should be highlighted. If there are no entries in the archive, the first individual that has a fitness above f_{min} is added. Moreover, if the number of archive entries is below max_{arch}, Eq. 2 is used with the number of archive entries instead of max_{arch}.

3.2   Selection Mechanism

The selection mechanism is important to shape how evolution will proceed, depending on the results obtained in a given generation. Our novelty approach has a customised selection mechanism which can switch between a fitness-based strategy and a hybrid mechanism that considers both fitness and novelty. It starts as a fitness guided evolution; however, that can change according to a decision rule, which is described as:

$$\begin{cases} change\_to\_fitness, & feasible_{inds} < T_{min} \\ change\_to\_hybrid, & feasible_{inds} > T_{max}, \end{cases}$$

where $feasible_{inds}$ is the number of individuals of the current generation that have a fitness above the threshold f_{min}, T_{min} is the threshold used to verify if evolution should be changed to fitness, and T_{max} is used to verify if it should be changed to hybrid.

In fitness guided evolution, the tournament selection is based on the fitness values of the candidate solutions, as in a standard EA. If hybrid evolution is chosen, it is necessary to compute the novelty of each selected individual, and perform a Pareto-based tournament selection, using the novelty and fitness of each selected individual as two different objectives to maximise.

The novelty computation process is inspired by Lehman and Stanley's work, as described in Eq. 1, with one small change: the k most similar images are considered from the set of the selected individuals and the archive, instead of considering the whole population and the archive. An example of this novelty computation is illustrated in Fig. 2. In Fig. 2, considering $k = 4$ and a tournament size of 5, the dashed lines denote the chosen individuals to compute novelty, and it is possible to see that from the 4 nearest individuals picked, 3 were chosen from the tournament while the remaining one was chosen from the archive.

At this stage, each selected individual has a fitness and novelty value, and there is the need to determine the winner of the tournament. This process is inspired by multi-objective EAs, namely the Pareto-based approaches, which select the best individuals based on their dominance or non-dominance when compared to other individuals [26]. In this work, the hybrid tournament selection determines the non-dominant solutions by comparing, among the selected individuals, on the basis of both fitness and novelty. After computing the set of non-dominant individuals, we have the so-called Pareto front. The tournament winner will be selected by randomly retrieving one of the solutions of the Pareto front.

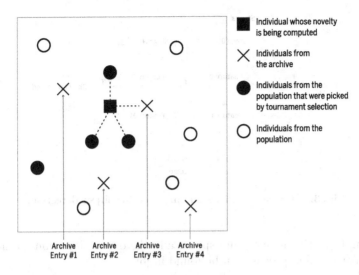

Individual whose novelty
is being computed

Individuals from
the archive

Individuals from the
population that were picked
by tournament selection

Individuals from the
population

Archive   Archive   Archive   Archive
Entry #1   Entry #2   Entry #3   Entry #4

Fig. 2. Novelty computation for an individual.

4 Evolving Figurative Images

One of the problems for which we decided to apply the proposed algorithm was the evolution of figurative images. In this problem, an expression-based approach is used to build the individuals' genotypes and the evolution is guided with the help of an external object classifier. Previous works tackled the evolution of singular objects such as faces, lips, leaves or flowers [6–8], and the evolution of several objects at the same time, attempting to create ambiguous images [9].

For this work, we used the geNeral purpOse expRession Based Evolutionary aRt Tool (norBErT), which is able to evolve figurative images using several evolution strategies [25]. It uses a tree representation to encode individuals and create images from those trees, using a rendering process which consists in generating an output value for each image pixel, which represents its colour.

For the fitness assignment task, an object detector is used to assess the images' quality. This object detector is a cascade classifier based on the work of Viola and Jones [27]. The architecture of this cascade classifier is shown in Fig. 3. The cascade classifier is composed of several stages and each stage tests the presence of a group of low level features (Local Binary Pattern features [28]).

The fitness function is built by obtaining internal information from the cascade classifier and combining them as follows:

$$f(x) = \sum_{i}^{\text{cstages}_x} (\text{stagedif}_x(i) \times i) + \text{cstages}_x \times 10, \qquad (3)$$

where cstages_x is the number of stages that an image x passed and stagedif_x is the difference between the score obtained in a given stage by an image x and the threshold necessary to pass that stage. The detailed description of the problem

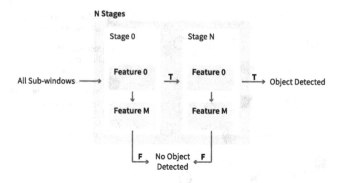

Fig. 3. Cascade classifier structure used for object detection.

of evolving figurative images, and specifically the object detection method and the classifier training process can be found in [6].

5   Evolving Context Free Design Grammars

In the last section we studied how novelty search mechanisms can be applied for the evolution of figurative images. Aiming at better understanding the impact of novelty mechanisms applied to a broader scope of evolutionary algorithms, we applied this kind of procedure to a different environment, which works as a proof of concept for the ability of the methodology to properly evolve solutions under different domains.

Taking into account the previous statements, we address the evolution of CFDGs [10], which, in simple words, are a powerful way of generating images through a compact set of rules (for full description check [10]). CFDGs are mapped into images by using a software tool named Context Free Art [29].

The evolutionary engine used for this task is an extension of the one discussed in [30,31], which has been generalised, in order to enable it to deal with every domain that can be represented by means of a formal grammar (in the Backus-Naur Form). It is thoroughly described in [32] and the application to the domain of CFDGs, along with a large set of comprehensive experiments, is detailed in [11].

The purpose of the engine is to evolve individuals expressed from grammar formulations. In essence, a grammar is formed by a 4-tuple: (V, Σ, R, S) where: V is a set of non-terminal symbols; Σ is a set of terminal symbols; R is a set of production rules that map from V to $(V \cup \Sigma)^*$; S is the initial symbol. Additionally, grammars can also be considered augmented, because we allow the specification of parameters in the calls to terminal and non-terminal symbols. One can also say that the grammars are non-deterministic, because it is possible to define the same non-terminal symbol more than once. When several production rules are applicable, one of them is randomly selected and the expansion proceeds. Figure 4 depicts an example of a CFDG and several images that can

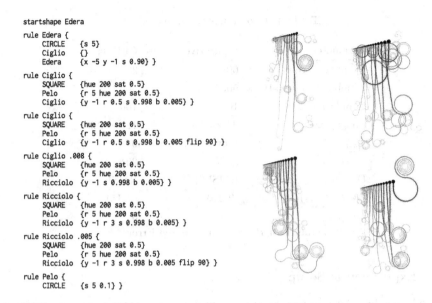

```
startshape Edera

rule Edera {
    CIRCLE    {s 5}
    Ciglio    {}
    Edera     {x -5 y -1 s 0.90} }

rule Ciglio {
    SQUARE    {hue 200 sat 0.5}
    Pelo      {r 5 hue 200 sat 0.5}
    Ciglio    {y -1 r 0.5 s 0.998 b 0.005} }

rule Ciglio {
    SQUARE    {hue 200 sat 0.5}
    Pelo      {r 5 hue 200 sat 0.5}
    Ciglio    {y -1 r 0.5 s 0.998 b 0.005 flip 90} }

rule Ciglio .008 {
    SQUARE    {hue 200 sat 0.5}
    Pelo      {r 5 hue 200 sat 0.5}
    Ricciolo  {y -1 s 0.998 b 0.005} }

rule Ricciolo {
    SQUARE    {hue 200 sat 0.5}
    Pelo      {r 5 hue 200 sat 0.5}
    Ricciolo  {y -1 r 3 s 0.998 b 0.005} }

rule Ricciolo .005 {
    SQUARE    {hue 200 sat 0.5}
    Pelo      {r 5 hue 200 sat 0.5}
    Ricciolo  {y -1 r 3 s 0.998 b 0.005 flip 90} }

rule Pelo {
    CIRCLE    {s 5 0.1} }
```

Fig. 4. CFDG adapted from www.contextfreeart.org/gallery/view.php?id=165 together with examples of images produced by its rendering.

result thanks to the probabilistic nature of its associated rendering algorithm. More information regarding representation and genetic operators for CFDGs can be found in [11,32].

To test the influence of novelty in evolution, we adapted the methodology described in Sect. 3 to this evolutionary engine. To assess the quality of each individual we use a combination of aesthetic measures: contrasting colours [11] and Bell Curve fit. Bell Curve fit [33] is based on the observation that fine-art works tend to exhibit a normal distribution of colour gradients. Thus, we calculate the deviation from the Gaussian normal distribution of colour gradients of the individuals. Contrasting colours counts the number of colours present in an image, discarding similar ones. Therefore, the engine aims at promoting the evolution of images that depict a wide range of colours, possessing a normal distribution of their gradients.

6 Results

In the current section, we describe the set of experiments that were conducted using the two different engines summarised in Sects. 4 and 5, respectively for the evolution of figurative images and CFDGs. In Sect. 6.1 we present the setup used for each problem, while Sect. 6.2 details the results. For each problem, we compare the performance of a standard fitness evolution approach and our proposed algorithm, which takes novelty into account.

Table 1. Experimental parameters.

Parameter	Figurative images	CFDGs
Number of runs	60	30
Number of generations	500	500
Population size	100	100
Image size	64×64	128×128
f_{min}	Flower detected	0.52
$dissim_{min}$	20 %	20 %
Dissimilarity metric	RMSE[a]	RMSE
(T_{min}, T_{max})	$(2, 15)$	$(2, 15)$
max_{arch}	5	5

[a]RMSE stands for Root Mean Squared Error

6.1 Experimental Setup

Table 1 specifies the parameters used to perform the experiments for evolving figurative art images and CFDGs. We performed 30 runs for the problem of evolving CFDGs and 60 runs in the evolution of figurative images.

The evolution of figurative images sometimes results in an empty archive. Therefore, a larger number of runs were needed to ensure a sufficient number of runs with non-empty archive. The entry condition for the archive is different in the two cases. In norBErT, classifiers are used to detect objects and therefore, a feasible solution would be one that contains an object. In the current section, we conducted tests aiming to get images that resemble flowers. In the case where CFDGs are evolved, a fitness threshold is defined for feasible solutions (in the current scenario, 0.52, which was empirically determined).

The remaining parameters, i.e., those specific to the evolution of figurative images and CFDGs can be found, respectively, in [25,32].

6.2 Results

The experiments serve as basis for comparing fitness-based and hybrid guided evolutions for figurative images and CFDGs. For both domains, we use the average fitness of the best individuals (Fig. 5), archive sizes (Fig. 6) and populational RMSE (Fig. 7) as comparison metrics. The evolution towards feasible zones of the search space – i.e., containing images recognized as flowers when evolving figurative images, or with a fitness above the considered threshold when evolving CFDGs – was observed in both problems. However, there is clearly a difference between the two problems in terms of difficulty. While in the evolution of CFDGs the f_{min} constant is determined empirically and based on subjective options, in the evolution of figurative images the constraint is more difficult to match, as images need to resemble an object. Figures 6 and 7 highlight this situation, as the archive sizes are higher in the CFDGs case and populations are more diverse.

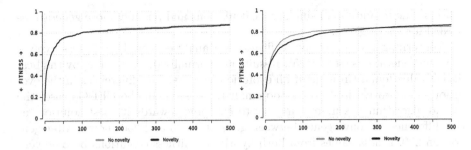

Fig. 5. Evolution of fitness across generations. On the left for figurative images (averages of 60 independent runs), on the right for CFDGs (averages of 30 independent runs).

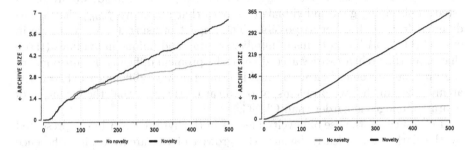

Fig. 6. Evolution of the archive size across generations. On the left for figurative images (averages of 60 independent runs), on the right for CFDGs (averages of 30 independent runs). Be aware of the difference in scale.

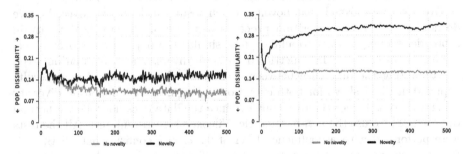

Fig. 7. Evolution of population dissimilarity across generations. On the left for figurative images (averages of 60 independent runs), on the right for CFDGs (averages of 30 independent runs).

This suggests that, when evolving CFDGs, feasible zones of the search space are more accessible.

Regardless of the difficulty of the problem, we aim to demonstrate that the proposed hybrid algorithm is able to outperform the fitness-only approach in terms of generating diverse images, with some cost associated in the evolutionary process, due to additional similarity calculations and possibly more generations needed for convergence.

Concerning the behaviour of fitness, we analysed how best fitness evolves over the generations (see Fig. 5). Results are normalised, by dividing by the best fitness value found throughout all evolutionary runs. The differences in the evolution of figurative images are too small to be seen. In the CFDGs case, the hybrid algorithm has more difficulty to converge towards its best solution, as the difference in convergence between generation 25 and 250 shows. After generation 250, fitness values from both hybrid and fitness evolutions become very close. This is due to the fact that evolving using a novelty metric blurs the search for the fittest individuals. The evolution process can thus move towards new areas, allowing us to get more unique individuals, at the cost of lower fitness values. Our main goal is not to maximize fitness *per se*, instead, we are interested in obtaining good individuals (the ones that are above f_{min}) that are as different from each other as possible. The impact of using a novelty metric in our hybrid algorithm in terms of number of good and unique individuals (those that enter the archive) varies according to the problem difficulty. Figure 6 makes the previous statement clear, by highlighting the huge difference between the archive sizes of the two problems, also indicating this way that the evolution of figurative images is harder than CFDGs.

In both cases, the use of novelty metrics in the hybrid algorithm, represented by the darker line, clearly favours the growth of the archive. The difference between fitness-based and hybrid evolution is larger in the case of the simpler problem, where it is possible to get, on average, roughly 360 additional diverse images per run when using the hybrid algorithm. In the case of evolving figurative images, the difference is noticeable but smaller (approximately 3 images).

We have established that novelty search results in larger archives, but we do not know how the algorithm proceeds in terms of phenotypic diversity in a broader scope. In order to analyse how similar the population is over generations, we used the RMSE metric to measure diversity at a population level, by computing the difference between every pair of individuals. As the results depicted in Fig. 7 show, for both evolutionary engines, the population diversity using novelty is higher than without it. Since the data does not follow a normal distribution, we use the non-parametric Wilcoxon signed-rank test. All the tests were performed with a confidence level of 95 %, and confirm that all the differences reported herein are statistically significant. Moreover, almost across all the generations, the difference between the population RMSE, with and without novelty is higher when evolving CFDGs. This can also be attributed to the fact that it is easier to evolve CFDGs than figurative images.

When evolving figurative images, diversity with novelty converges faster (approximately to 0.14) than when evolving CFDGs, which seems to increase till approximately 0.31. This is expected due to the nature of the problems and to the parameters that were chosen. While evolving figurative images, tournament changes from fitness only to hybrid are much more frequent than when evolving CFDGs. That also explains why at the beginning the diversity of CFDGs decreases. Initially, it is necessary to generate the minimum number of feasible

Fig. 8. Archive of a run evolving figurative images guided only by fitness.

Fig. 9. Archive (sampled) of a run evolving figurative images with the hybrid method.

Fig. 10. Archive of a run evolving CFDGs guided only by fitness.

solutions (CFDGs superior to a defined threshold of fitness), sacrificing diversity. Until the archive has at least one entry, the two methods are practically identical. Then, after changing to hybrid evolution, the novelty metrics promote an increase in population diversity, which is also noticeable by the growth of the archive, depicted in Fig. 6.

We consider it essential to analyse the results from two different perspectives: the computational one and the human one. While the former is covered by the RMSE-based analysis, the latter was explored through a subjective human analysis. Overall, we can conclude that not only the number of resulting images in the archive is larger in the cases when we use the hybrid algorithm, but also that the differences between them are more visually noticeable. This is confirmed by the archive images of a single representative run starting from the same initial population of images, using both approaches, for the problem of evolving figurative images (Figs. 8 and 9) and for evolving CFDGs (Figs. 10 and 11).

Fig. 11. Archive (sampled) of a run evolving CFDGs with the hybrid method.

7   Conclusions and Future Work

In this paper we proposed a new novelty search approach within the context of evolutionary art, and applied it to the evolution of figurative images and CFDGs, using two different evolutionary engines. Our approach is an hybrid algorithm with adapted evolution, in the sense that it changes between fitness based and hybrid evolution, taking into account fitness and novelty as two different objectives to maximise. Tests were designed to compare the performance of our novelty search approach with canonical fitness based evolution.

The experimental results show that our algorithm is able to promote the discovery of a wide set of phenotypically diverse and fit solutions to the problems considered, outperforming fitness based evolution in terms of the number and the diversity of the generated solutions.

Next steps to this work include the study of the conditions used to change the evolution method. At this moment, the method depends on two user-defined parameters. Choosing the best parameter settings requires some insight from the user, which may not be easy to obtain. Domain knowledge may be needed to understand what are the best parameter settings for each problem. We believe that a self-adaptive mechanism that automatically detects when to change the evolution method could be beneficial to this work. Additionally, the application of these techniques to other types of problems, outside the scope of evolutionary art, is also under way.

Acknowledgments. The project ConCreTe acknowledges the financial support of the Future and Emerging Technologies (FET) programme within the Seventh Framework Programme for Research of the European Commission, under FET grant number 611733. This research is also partially funded by: Fundação para a Ciência e Tecnologia (FCT), Portugal, under the grant SFRH/BD/90968/2012. The authors also acknowledge the feedback provided by the blind reviewers of this paper.

References

1. Boden, M.A.: The Creative Mind: Myths and Mechanisms. Psychology Press, New York (2004)
2. Kowaliw, T., Dorin, A., McCormack, J.: Promoting creative design in interactive evolutionary computation. IEEE Trans. Evol. Comput. 16(4), 523 (2012)
3. Sims, K.: Artificial evolution for computer graphics. ACM Comput. Graph. 25, 319–328 (1991)
4. Machado, P., Cardoso, A.: All the truth about NEvAr. Appl. Intel. 16(2), 101–119 (2002). Special Issue on Creative Systems
5. McCormack, J.: Facing the future: evolutionary possibilities for human-machine creativity. In: Romero, J., Machado, P. (eds.) The Art of Artificial Evolution. Natural Computing Series, pp. 417–451. Springer, Heidelberg (2008)
6. Correia, J., Machado, P., Romero, J., Carballal, A.: Evolving figurative images using expression-based evolutionary art. In: Proceedings of the Fourth International Conference on Computational Creativity (ICCC), pp. 24–31 (2013)
7. Machado, P., Correia, J., Romero, J.: Expression-based evolution of faces. In: Machado, P., Romero, J., Carballal, A. (eds.) EvoMUSART 2012. LNCS, vol. 7247, pp. 187–198. Springer, Heidelberg (2012)
8. Machado, P., Correia, J., Romero, J.: Improving face detection. In: Moraglio, A., Silva, S., Krawiec, K., Machado, P., Cotta, C. (eds.) EuroGP 2012. LNCS, vol. 7244, pp. 73–84. Springer, Heidelberg (2012)
9. Machado, P., Vinhas, A., Correia, J.A., Ekárt, A.: Evolving ambiguous images. In: Proceedings of the 24th International Conference on Artificial Intelligence, IJCAI 2015, pp. 2473–2479. AAAI Press (2015)
10. Horigan, J., Lentczner, M.: Context Free Design Grammar version 2 syntax (2015). http://www.contextfreeart.org/mediawiki/index.php/Version_2_Syntax
11. Machado, P., Correia, J., Assunção, F.: Graph-based evolutionary art. In: Gandomi, A., Alavi, A.H., Ryan, C. (eds.) Handbook of Genetic Programming Applications. Springer, Heidelberg (2015)
12. Krcah, P., Toropila, D.: Combination of novelty search and fitness-based search applied to robot body-brain co-evolution. In: Czech-Japan Seminar on Data Analysis and Decision Making in Service Science, pp. 1–6 (2010)
13. Methenitis, G., Hennes, D., Izzo, D., Visser, A.: Novelty search for soft robotic space exploration. In: Proceedings of the 2015 on Genetic and Evolutionary Computation Conference, GECCO 2015, pp. 193–200. ACM, New York (2015)
14. Mouret, J.-B.: Novelty-based multiobjectivization. In: Doncieux, S., Bredèche, N., Mouret, J.-B. (eds.) New Horizons in Evolutionary Robotics. SCI, vol. 341, pp. 139–154. Springer, Heidelberg (2011)
15. Liapis, A., Yannakakis, G.N., Togelius, J.: Sentient sketchbook: computer-aided game level authoring. In: FDG, pp. 213–220 (2013)
16. Secretan, J., Beato, N., D'Ambrosio, D.B., Rodriguez, A., Campbell, A., Folsom-Kovarik, J.T., Stanley, K.O.: Picbreeder: a case study in collaborative evolutionary exploration of design space. Evol. Comput. 19(3), 373–403 (2011)
17. Lehman, J., Stanley, K.O.: Exploiting open-endedness to solve problems through the search for novelty. In: Proceedings of the Eleventh International Conference on Artificial Life (ALIFE XI). MIT Press, Cambridge (2008)
18. Saunders, R., Gero, J.S.: The digital clockwork muse: a computational model of aesthetic evolution. Proc. AISB 1, 12–21 (2001)

19. Silberschatz, A., Tuzhilin, A.: What makes patterns interesting in knowledge discovery systems. IEEE Trans. Knowl. Data Eng. **8**(6), 970–974 (1996)
20. Kohonen, T.: Self-Organization and Associative Memory, 3rd edn. Springer, New York (1989)
21. Lehman, J., Stanley, K.O.: Revising the evolutionary computation abstraction: minimal criteria novelty search. In: Proceedings of the 12th Annual Conference on Genetic and Evolutionary Computation, pp. 103–110. ACM (2010)
22. Liapis, A., Yannakakis, G., Togelius, J.: Enhancements to constrained novelty search: two-population novelty search for generating game content. In: Proceedings of Genetic and Evolutionary Computation Conference (2013)
23. Kimbrough, S.O., Koehler, G.J., Lu, M., Wood, D.H.: On a feasible-infeasible two-population (FI-2Pop) genetic algorithm for constrained optimization: distance tracing and no free lunch. Eur. J. Oper. Res. **190**(2), 310–327 (2008)
24. Cuccu, G., Gomez, F.: When novelty is not enough. In: Di Chio, C., et al. (eds.) EvoApplications 2011, Part I. LNCS, vol. 6624, pp. 234–243. Springer, Heidelberg (2011)
25. Vinhas, A.: Novelty and figurative expression-based evolutionary art. Master's thesis, Department of Informatic Engineering, Faculty of Sciences and Technology, University of Coimbra, July 2015
26. Fonseca, C.M., Fleming, P.J.: An overview of evolutionary algorithms in multiobjective optimization. Evol. Comput. **3**(1), 1–16 (1995)
27. Viola, P., Jones, M.: Rapid object detection using a boosted cascade of simple features. In: Proceedings of the 2001 IEEE Computer Society Conference on Computer Vision and Pattern Recognition, vol. 1, p. 511 (2001)
28. Ojala, T., Pietikäinen, M., Harwood, D.: A comparative study of texture measures with classification based on feature distributions. Pattern Recogn. **29**(1), 51–59 (1996)
29. Horigan, J., Lentczner, M.: Context Free (2014). http://www.contextfreeart.org/
30. Machado, P., Nunes, H.: A step towards the evolution of visual languages. In: First International Conference on Computational Creativity, Lisbon, Portugal (2010)
31. Machado, P., Nunes, H., Romero, J.: Graph-based evolution of visual languages. In: Di Chio, C., et al. (eds.) EvoApplications 2010, Part II. LNCS, vol. 6025, pp. 271–280. Springer, Heidelberg (2010)
32. Assunção, F.: Grammar based evolutionary design. Master's thesis, Department of Informatic Engineering, Faculty of Sciences and Technology, University of Coimbra, July 2015
33. Ross, B.J., Ralph, W., Zong, H.: Evolutionary image synthesis using a model of aesthetics. In: IEEE Congress on Evolutionary Computation, CEC 2006, pp.1087–1094. IEEE (2006)

Author Index

Printed in the United States
By Bookmasters